Genetic and Evolutionary Computation

Genetic and Evolutionary Computation
Medical Applications

Edited by

Stephen L. Smith

Department of Electronics, University of York, York, UK

Stefano Cagnoni

Dipartimento di Ingegneria dell'Informazione, Università degli Studi di Parma, Parma, Italy

A John Wiley and Sons, Ltd., Publication

This edition first published 2011
© 2011 John Wiley & Sons, Ltd
Chapter 5.3 2011 © the Crown in right of Canada

Registered office
John Wiley & Sons Ltd, The Atrium, Southern Gate, Chichester, West Sussex, PO19 8SQ,
United Kingdom

For details of our global editorial offices, for customer services and for information about how
to apply for permission to reuse the copyright material in this book please see our website at
www.wiley.com.

Library of Congress Cataloguing-in-Publication Data

Genetic and evolutionary computation : medical applications / edited by Stephen L. Smith,
Stefano Cagnoni.
    p. ; cm.
    Includes index.
    ISBN 978-0-470-74813-8 (cloth)
    1. Medicine–Data processing.   2. Genetic programming (Computer science)
3. Evolutionary programming (Computer science)   I. Smith, Stephen L., 1962–
II. Cagnoni, Stefano, 1961–
    [DNLM: 1. Computational Biology–methods. 2. Models, Genetic. 3. Algorithms.
4. Evolution, Molecular. QU 26.5]
    R858.G457 2010
    610.285–dc22

                                                                    2010025421

A catalogue record for this book is available from the British Library.

Print ISBN: 9780470748138
ePDF ISBN: 9780470973141
oBOOK ISBN: 9780470973134

Set in 10/12.5pt Palatino by Aptara Inc., New Delhi, India.
Printed and bound in Singapore by Markono Print Media Pte Ltd.

Contents

About the Editors

Stephen L. Smith received a BSc in Computer Science and then an MSc and PhD in Electronic Engineering from the University of Kent, UK. He is currently a senior lecturer in the Department of Electronics at the University of York, UK.

Stephen's main research interests are in developing novel representations of evolutionary algorithms, particularly with application to problems in medicine. His work is currently centred on the diagnosis of neurological dysfunction and analysis of mammograms. The former is currently undergoing clinical trials and being registered for a patent.

Stephen was program chair for the Euromicro Workshop on Medical Informatics, program chair and local organizer for the Sixth International Workshop on Information Processing in Cells and Tissues (IPCAT), and guest editor for the subsequent special issue of *BioSystems* journal. Most recently he was tutorial chair for the IEEE Congress on Evolutionary Computation (CEC) in 2009. Stephen is Associate Editor for *Genetic Programming and Evolvable Machines* journal, the *Journal of Artificial Evolution and Applications*, and a member of the editorial board for the *International Journal of Computers in Healthcare*. Along with Stefano Cagnoni, Stephen co-founded the annual GECCO Workshop on Medical Applications of Genetic and Evolutionary Computation which is now in its seventh year.

Stefano Cagnoni graduated in Electronic Engineering at the University of Florence in 1988 where he was a PhD student and a post-doc until 1997. In 1994 he was a visiting scientist at the Whitaker College Biomedical Imaging and Computation Laboratory at the Massachusetts Institute of Technology. Since 1997 he has been with the University of Parma, where he has been Associate Professor since 2004.

Stefano's main basic research interests concern soft computing, with particular regard to evolutionary computation and computer vision. As concerns applied research, the main topics of his research are the application of the above-mentioned techniques to problems in computer vision, pattern recognition and robotics. Recent research grants regard: co-management of a project funded by Italian Railway Network Society (RFI) aimed at developing an automatic inspection system for train pantographs; a 'Marie Curie' grant, for a four-year research training project in Medical Imaging using Bio-Inspired and Soft Computing; a grant from 'Compagnia di S. Paolo' on 'Bioinformatic and experimental dissection of the signalling pathways underlying dendritic spine function'.

Stefano was Editor-in-chief of the *Journal of Artificial Evolution and Applications* from 2007 to 2009 and has been chairman of EvoIASP since 1999 (an event dedicated to evolutionary computation for image analysis and signal processing). Stefano is also a co-editor of special issues of journals dedicated to Evolutionary Computation for Image Analysis and Signal Processing and has been a reviewer for international journals and a member of the committees of several conferences. He is a member of the Advisory Board of Perada, the UE Coordination Action on Pervasive Adaptation and has recently been awarded the 'Evostar 2009 Award', in recognition of his most outstanding contribution to Evolutionary Computation.

List of Contributors

Editors

Stephen L. Smith
Department of Electronics, University of York, York, UK

Stefano Cagnoni
Dipartimento di Ingegneria dell'Informazione, Università degli Studi di Parma, Parma, Italy

Contributors

Linda El Alaoui
Département de Mathématiques, Institut Galilée, Villetaneuse, France

Barbara G. Beckerman
Oak Ridge National Laboratory, Oak Ridge, USA

Vitoantonio Bevilacqua
Department of Electrical and Electronics, Polytechnic of Bari, Bari, Italy
e.B.I.S. s.r.l. (electronic Business In Security), Spin-Off of Polytechnic of Bari, Bari, Italy

Alan J. Barton
National Research Council Canada, Ottawa, Ontario, Canada

Stefano Cagnoni
Dipartimento di Ingegneria dell'Informazione, Università degli Studi di Parma, Parma, Italy

Simona Cambò,
Department of Electrical and Electronics, Polytechnic of Bari, Bari, Italy

Lucia Cariello
Department of Electrical and Electronics, Polytechnic of Bari, Bari, Italy
e.B.I.S. s.r.l. (electronic Business In Security), Spin-Off of Polytechnic of Bari, Bari, Italy

Santiago E. Conant-Pablos
Centro de Computación Inteligente y Robótica, Tecnológico de Monterrey, Monterrey, México

Crispin Cooper
Department of Electronics, University of York, York, UK

Domenico Daleno
Department of Electrical and Electronics, Polytechnic of Bari, Bari, Italy
e.B.I.S. s.r.l. (electronic Business In Security), Spin-Off of Polytechnic of Bari, Bari, Italy

Laurent Dumas
Laboratoire Jacques-Louis Lions, Université Pierre et Marie Curie, Paris, France

Ghassan Hamarneh
Medical Image Analysis Lab, Simon Fraser University, Burnaby, Canada

Rolando R. Hernández-Cisneros,
Centro de Computación Inteligente y Robótica, Tecnológico de Monterrey, Monterrey, México

David M. Howard
Department of Electronics, University of York, York, UK

Michael A. Lones
Department of Electronics, University of York, York, UK

D. Lungeanu
Faculty of Mathematics and Informatics, West University Of Timisoara, Timiş, Romania

Giuseppe Mastronardi
Department of Electrical and Electronics, Polytechnic of Bari, Bari, Italy
e.B.I.S. s.r.l. (electronic Business In Security), Spin-Off of Polytechnic of Bari, Bari, Italy

Chris McIntosh
Medical Image Analysis Lab, Simon Fraser University, Burnaby, Canada

Robert Orchard
National Research Council Canada, Ottawa, Ontario, Canada

Robert M. Patton
Oak Ridge National Laboratory, Oak Ridge, USA

Thomas E. Potok
Oak Ridge National Laboratory, Oak Ridge, USA

Hugo Terashima-Marín
Centro de Computación Inteligente y Robótica, Tecnológico de Monterrey, Monterrey, México

Andy M. Tyrrell
Department of Electronics, University of York, York, UK

Stephen L. Smith
Department of Electronics, University of York, York, UK

Julio J. Valdés
National Research Council Canada, Ottawa, Ontario, Canada

Leonardo Vanneschi
Dipartimento di Informatica, Sistemistica e Comunicazione, Università di Milano-Bicocca, Milan, Italy

Daniela Zaharie
Faculty of Mathematics and Informatics, West University Of Timisoara, Timiş, Romania

Flavia Zamfirache
Faculty of Mathematics and Informatics, West University Of Timisoara, Timiş, Romania

1

Introduction

Genetic and evolutionary computation (GEC) is now attracting considerable interest, since the first algorithms were developed some 30 years ago. Although often regarded as a theoretical pursuit, research and development of a wide range of real-world applications of GEC has long been evident at conferences and in the scientific literature. Medicine and healthcare is no exception and the challenge, and worthy aim, has motivated many to apply GEC to a wide range of clinical problems.

The aim of this book is to provide an overview of the range of GEC techniques being applied to medicine and healthcare, in a context that is supportive not only for existing GEC practitioners, but also for those from other disciplines, particularly health professionals. This encompasses doctors, consultants and other clinicians, as well as those who act in a technical role in the health industry, such as medical physicists, technicians and those who have an interest in learning more with a view to implementing systems or just understanding them better. Consequently, a concise introduction to genetic and evolutionary computation is required, and this has been provided in Chapter 2 for those readers who are not familiar with the more common paradigms of GEC.

It was also felt important that an overview of recent work should be reported as concisely and fully as practically possible, and this has been provided in Chapter 3. The problem with any review is that, despite the best efforts of the author, it is outdated, incomplete and unbalanced as soon as it has been published. It is also impossible to know which papers are going to be of future significance, for the individual or community as a whole, regardless of the subject, author or source of publication. There is also the risk

Genetic and Evolutionary Computation: Medical Applications   Edited by Stephen L. Smith and Stefano Cagnoni
© 2011 John Wiley & Sons, Ltd

that the review becomes too cumbersome to maintain the reader's interest, comprising an endless list of summaries with little structure or context. For these reasons, this review has adopted two guiding principles. Firstly, it is limited to the last five years of publications, as it is felt that this will encompass very recent approaches and yet previous work of merit will have been refined, extended or combined with other techniques and reported in more recent publications. Secondly, no distinction has been made on the grounds of source of publication, whether it is a journal, conference or workshop presentation. It is anticipated, however, that all papers will have been peer reviewed to provide some level of confidence in the work presented. The overriding aim of the review is to stimulate thought on how techniques investigated to date may be used to the reader's advantage.

The main component of this book is a set of nine case examples on the application of GEC to different areas of medicine, which have been grouped into three chapters covering medical imaging, the analysis of medical data sets, and medical modelling, diagnosis and treatment. This is by no means a representative selection, but one that conveys the breadth of techniques employed. The source of these contributions has been the Genetic and Evolutionary Computation Conference (GECCO) Workshop on Medical Applications of Genetic and Evolutionary Computation (MedGEC), which has been part of GECCO since 2005. The Workshop has provided a valuable venue for reporting work, often in early stages of development, but which has then matured to be published in GEC journals – notable examples are special issues in the journal *Genetic Programming and Evolvable Machines* and the *Journal of Artificial Evolution and Applications*.

The final chapter of this book will then consider the future of medical applications of GEC, the opportunities, challenges and rewards that practitioners face.

2

Evolutionary Computation: A Brief Overview

Stefano Cagnoni[1] and Leonardo Vanneschi[2]

[1]*Dipartimento di Ingegneria dell'Informazione, Università degli Studi di Parma, Parma, Italy*
[2]*Dipartimento di Informatica, Sistemistica e Comunicazione, Università di Milano-Bicocca, Milan, Italy*

2.1 Introduction

Virtually every event occurring in everybody's life is subject to chance. It is always amazing to realize how much our life could have been different, had not some apparently trivial and purely accidental events happened at specific moments. The feeling of being 'in the hands of Fate', and the way humans have dealt with this, have had a strong impact on the course of human history and civilization.

The rational nature of humans and the self-consciousness of their ability to be more and more the master of their own future has turned mankind's attitude towards chance from the primitive populations' passive acceptance, to the attempt to make forecasts based on the study of the stars or other natural phenomena, to the illuministic rejection of any influence of chance on human life, to the present, rational/scientific approach by which chance is seen as a component of the dynamics of life which is to be studied, biased and possibly exploited to mankind's advantage. In this regard, nature has often

Genetic and Evolutionary Computation: Medical Applications Edited by Stephen L. Smith and Stefano Cagnoni
© 2011 John Wiley & Sons, Ltd

offered brilliant examples about how chance can be turned into a constructive process, essential for the survival of living beings.

Among such examples, evolution is definitely one of the most fascinating. According to darwinian theory, a population can evolve and improve its survival ability just as a result of its 'natural' response to random events (or *mutations*). In fact, better individuals live longer and have higher chances of producing offspring, up to the point at which their genetic characters become part of the population's specific features. Evolution is therefore a striking example of how chance can become the driving engine for vital processes, when it is associated with suitable, possibly spontaneous, feedbacks that bias its outcomes. This is the reason why evolutionary processes, as has happened for many natural phenomena, have been emulated by computers, in order to translate their principles into powerful learning and design tools.

Evolutionary computation (EC) consists of a set of techniques which emulate darwinian evolution and its dynamics to accomplish tasks like function optimization, learning, self-adaption, and autonomous design. In this chapter we briefly describe the main concepts underlying the most common EC paradigms. For those who are interested in deepening their knowledge of EC, or in participating in the main EC events worldwide, a list of useful links is provided in the appendix to this book.

2.2 Evolutionary Computation Paradigms

The term *natural evolution* is generally used to indicate the process that has transformed the community of living beings populating Earth from a set of simple unicellular organisms to a huge variety of animal species, well integrated with the surrounding environment. Darwin [1] identified a small set of essential elements which rule evolution by natural selection: reproduction of individuals, variation phenomena that affect the likelihood of survival of individuals, inheritance of many of the parents' features by offspring in reproduction, and the presence of a finite amount of resources causing competition for survival between individuals. These simple features – reproduction, likelihood of survival, variation, inheritance and competition – are the bricks that build the simple model of evolution that inspired the computational intelligence technique known as *evolutionary algorithms* (EAs). EAs work by defining a goal in the form of a quality criterion and then use this criterion to measure and compare a set, or *population*, of solution candidates. These potential solutions (*individuals* in the population) are generally data structures. The process on which EAs are based is an iterative stepwise refinement of the quality of these structures. To refine individuals, this process uses a set of operators directly inspired by natural evolution. These operators are, basically, *selection* and *variation*. Selection is the process that allows the best individuals to be chosen for mating, following Darwin's idea of likelihood of survival and competition. The two main variation operators in EAs, as in nature, are *mutation* and sexual reproduction (more often called recombination or *crossover*). Mutation changes a small part of an individual's structure while crossover exchanges some features of a set of (usually two) individuals to create offspring that are a combination of their parents. In some senses, mutation can be thought of as an *innovation* operator, since it introduces some brand new genetic material into the population. On the other hand, crossover is a *conservation* operator, in the sense that it uses the genetic material that is already present in the population, attempting to redistribute it, to produce better-quality individuals. In EAs, the quality of the individuals composing populations, or their likelihood of survival, is often called *fitness* and is usually measured by an algebraic function called the fitness function. The full expression of an individual, whose fitness is directly measurable, is often called *phenotype*, in opposition to the term *genotype*, which indicates the syntactical structure of which the individual is the decoding, just as living beings are the decoding of their DNA code. Clearly, in EAs as in nature, inheritance, mutation, and recombination act on the genotype, while selection acts on the phenotype: physical features are handed on from parents to offspring, while individuals survive the surrounding environment as a function of their ability to adapt to it. Many different kinds of EAs have been developed over the years. The feature that, more than any other, distinguishes the different paradigms of EAs is the way in which individuals are represented (this will be addressed as the *representation* problem in the following) and, consequently, the definitions of the genetic operators working on them. The art of choosing an appropriate representation and an appropriate set of operators is often a matter of experience and intuition.

2.2.1 Genetic Algorithms

Genetic algorithms (GAs) are one of the oldest and best known kinds of EA. They were invented by Holland in the early 1970s [9], and successfully applied to a wide variety of real-world problems, such as combinatorial optimization or learning tasks. Individuals are encoded as *fixed-length strings of characters*. The first step in the design of a GA is the definition of the set of n possible characters that can be used to construct the strings and the string length

L. The search space (i.e., the set of all possible individuals) is thus composed by n^L different strings. A frequently used particular case (often called the *canonical genetic algorithm*) is that in which the possible characters are just 0 and 1, and thus the search space size is 2^L.

The GA process starts with the generation of an initial population, composed of a set of randomly chosen individuals belonging to the search space. The size p of the population is usually fixed at the beginning of the GA process and does not change during evolution. Normally, $p \ll n^L$. After creating the initial population, the process enters a loop, where the following steps are performed at each iteration (iterations of the GA process are usually called *generations*): (i) each string is evaluated and assigned a fitness value. (ii) selection is performed: one individual is selected from the current population and inserted into an *intermediate population* (also called a *mating pool*). Without removing individuals selected in the previous steps, the process is repeated p times, so that, at the end of the selection process, the intermediate population is composed of p individuals. (iii) variation is performed: reproduction, crossover, and mutation are applied to the individuals of the intermediate population to create the *next population*. The process ends on the basis of a *termination criterion*. The most commonly used termination criteria are: (1) at least one individual in the current population has a satisfactory fitness value, or (2) a prefixed number of generations has been executed.

Many selection algorithms have been developed to build the intermediate population from the current one. All of them are based on the fitness of the individuals: individuals with higher fitness values have a higher probability of being chosen for mating. This does not ensure that the best individual is included in the next generation. If survival of the fittest individual (or of the n fittest individuals) is enforced, the algorithm is said to be 'elitist'.

As regards variation operators, reproduction consists simply of copying an individual from the intermediate population into the next one. Crossover in GAs may occur following many possible algorithms: the most common one is called *one-point crossover* [9]. Its behavior is described in Figure 2.1.

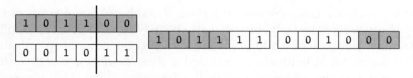

Figure 2.1   An example of GA crossover

1	1	0	0	1	0

1	1	1	0	1	0

Figure 2.2 An example of GA mutation

Many mutation algorithms have also been developed. The most commonly used is called *point mutation* [9], and acts by flipping single bits with a preset probability. Its behavior is shown in Figure 2.2.

The binary representation is virtually universal, and is particularly useful in case of mixed-type phenotypes, i.e., where the binary string may encode attributes (*genes*) of different type. However, the choice of the representation by which an individual is encoded has a significant impact on the features of the function which is to be optimized (the so-called *fitness landscape*). Therefore, specific representations and operators have been developed for integer- or real-valued phenotypes, permutations, etc.

2.2.2 Evolution Strategies

Evolution strategies (ESs) owe their origin to the work of Rechenberg and Schwefel [2,3,26]. Individuals are represented as real-valued vectors and the genetic operators working on them are selection, recombination, and mutation. ESs went through a long period of stepwise development after the formulation of the basic ideas. The first version was called *two-membered* ES, or (1+1)-ES, by Rechenberg [23]. It is based upon a 'population' consisting of one parent and one descendant, created by adding normally distributed random numbers to the parent. The best of both individuals then serves as the ancestor of the following iteration/generation. Indeed, the (1+1)-ES does not really use the population principle. To introduce this concept into the algorithm, Rechenberg proposed the *multimembered* ES, or (μ+1)-ES, where $\mu > 1$ parents can participate in the generation of one new individual. With the introduction of μ parents, instead of only one, it is possible to emulate sexual reproduction. The selection operator removes the least fit individual, may it be the offspring or one of the parents, from the population before the next generation starts producing new offspring. The mutation operator is still present, and it operates as in the (1+1)-ES. The next generalization step has been taken by Schwefel [27,28], with the introduction of the ($\mu + \lambda$)-ES and the (μ, λ)-ES. As the name ($\mu + \lambda$)-ES suggests, μ parents produce λ offspring, which are then reduced again to μ individuals making up the next generation. Selection operates on the joint set of parents and offspring. Thus, parents survive until they are superseded by better offspring. On the other hand, in the (μ, λ)-ES, only offspring undergo selection and can therefore be included in the next generation, i.e., the lifetime of every individual is limited to one generation.

2.2.3 Evolutionary Programming

Evolutionary programming (EP) was developed by Fogel, Owens, and Walsh [7]. Its process involves three steps (repeated until a threshold number of iterations is exceeded or an adequate solution is obtained): (1) choose an initial population of trial solutions at random; (2) replicate each solution into a new population. Each of these offspring solutions is mutated according to a distribution of mutation types, ranging from minor to extreme, with a continuum of mutation types in between. The severity of mutation is judged based upon the functional change imposed on the parents; (3) each offspring solution is assessed by computing its fitness. Typically, a stochastic tournament is held to determine N solutions to be retained, although this is occasionally performed deterministically. The main differences between EP and GAs are: (i) in EP there is no constraint on the representation. Since individuals can also be represented as computer programs, or, more frequently, as *finite state automata* (FSA) [7], EP can be considered as a forefather of genetic programming; (ii) in EP, the mutation operation simply changes aspects of the solution according to a statistical distribution, which weights minor variations in the behavior of the offspring as highly probable and substantial variations as increasingly unlikely. Furthermore, the severity of mutations is often reduced as the global optimum is approached; (iii) EP is an abstraction of evolution at the level of reproductive populations (i.e., species) and thus no recombination mechanism (such as crossover) is typically used, because recombination does not occur between species.

2.2.4 Genetic Programming

GAs are capable of solving many problems, simple enough to allow solid theoretical studies. Nevertheless, GAs' fixed-length string representation of individuals is difficult, unnatural, and constraining for a wide set of applications. This lack of representation power (already recognized in [2–4]) is overcome by *genetic programming* (GP) [12], which basically works as GAs, with the major difference that individuals to be evolved are not fixed-length strings of characters but, generally speaking, *computer programs*. The fitness

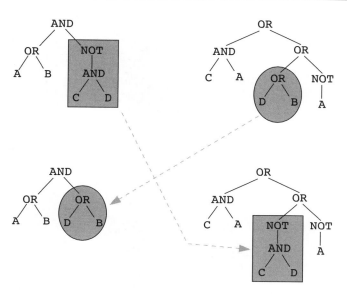

Figure 2.3 An example of standard GP crossover

of a program is usually calculated by running it one or more times with a variety of inputs (known as *fitness cases*) and seeing how close the program outputs are to some desired results. Programs can be represented as syntactic trees (for instance, the tree on the top left of Figure 2.3 encodes the Boolean function '(A OR B) AND NOT (C AND D)'), lines of code, expressions in prefix or postfix notation, strings of variable length, etc. New programs are produced either by mutation or by crossover. For tree-based GP, standard crossover [12] independently selects one random point in each parent and swaps the underlying subtrees, as shown in Figure 2.3.

Standard mutation [12] operates on only one parental program, choosing a point at random and replacing the underlying subtree with a randomly generated tree, as represented in Figure 2.4.

The principle is similar to that underlying GAs, but the important property must be respected that individuals resulting from applications of the genetic operators must be *syntactically valid* programs (property of syntactic *closure*) to ensure that any random combination of 'genes' encodes a valid function. The tree representation of individuals guarantees this property, and this is the main reason why tree representations of programs are most commonly used in GP [12].

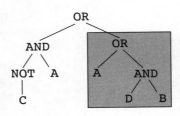

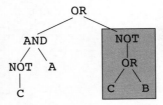

Figure 2.4 An example of standard GP mutation

2.2.5 *Other Evolutionary Techniques*

Besides the techniques we have just introduced, which make up the actual core of EC, there are several other techniques which can be considered either variants of the main paradigms, or nature- or artificial life-inspired methods loosely related to EC techniques, or, finally, techniques which are rooted into other more traditional AI techniques, to which evolutionary concepts are applied.

This is the case, for example, with *grammatical evolution* [25], a variant of GP based on the Backus–Naur representation of grammars in which individuals, encoded as a series of integers, can be decoded into full programs in any programming language by applying the production rules which define the syntax of the programming language.

An example of nature-inspired methods which are only loosely related to EC but, for historical reasons, are being studied within the EC community is the set of methods usually referred to as *swarm intelligence*. Among these, the most relevant are *ant colony optimization* (ACO) [5,6] and *particle swarm optimization* (PSO) [11,17]. These techniques translate the behavior of different animal species (in particular, ants and birds in search of food) into meta-heuristics for optimization. The region in space where these 'virtual animals' move is represented by the search space of the function to be optimized, while the quantity of food which can be found at a certain point is represented by the value of the fitness function at that point.

ACO's relevance is related both to being the first method of this group to be introduced and to being widely and successfully used in several applications. The model reflects what happens in nature when ants move (initially randomly) in search of food, releasing pheromone, which slowly tends to evaporate. If an ant finds food, it goes back to the nest to deposit it, 'marking' the path with pheromone. Since ants tend to follow pheromone traces during their search, over time, pheromone tends to accumulate along the paths which lead to food. The shorter the path, the more the pheromone trace is reinforced by ants which keep following it, as well as by lower evaporation. This way, over time, pheromone mostly accumulates along the shortest path, which becomes clearly marked for all ants to follow. Evaporation makes it possible for this process to be dynamical, so that the shortest path may change with the availability of food.

PSO, instead, is based on the emulation of the behavior of flocks of birds or schools of fish in search of food. The motion of a swarm element has: (i) a random component; (ii) a component based on local information which the element acquires directly and attracts it towards a local optimum it has found, as a bird flies around places where it has found food; and (iii) a component deriving from the experience of the whole swarm, which tends to attract it towards the best point (the point where most food has been found by the swarm) up to now. The emulation of swarm behavior leads to methods which are usually quite efficient at finding good sub-optimal points in the fitness landscape, as well as being able to deal with similar efficiency also with dynamically changing environments.

Finally, to close this limited review of EC or EC-like techniques, *learning classifier systems* (LCSs) [10,31] consist of a population of rules (which are binary in the simplest case, but can use more complex representations such as, for example, neural networks) of which a genetic algorithm alters and selects the best ones. Instead of using a fitness function, rule utility is decided by a reinforcement learning technique.

2.2.6 Theory of Evolutionary Algorithms

After reading this introduction to EAs one question may come as natural: why should EAs work at all? Or better: why should they build solutions of better and better fitness quality? And why should they find a solution that is satisfactory for a given problem? Or even better: what is the probability of improving the fitness quality of solutions along with generations? What is the probability of finding a satisfactory solution to a given problem? The attempt to answer these questions has been one of the main research activities in evolutionary computation since its early years. Being able to answer the above questions surely implies a deep understanding of what happens inside an EA population through generations, which is only possible by means of precise mathematical models. The most studied and best known classes of models in EAs are commonly known as *schema theories*. They are tightly related to the *building block hypothesis*, i.e., the idea that fractions of individuals exist (the building blocks) that improve the fitness of individuals which include them. For this reason, according to the building block hypothesis, individuals containing these substrings are often selected for mating, enabling building blocks to multiply their presence inside the population. Following this idea, a *schema* is a set of individuals in the search space sharing some syntactic feature. The general goal of schema theories is to provide information about the variation of some properties of individuals of a population belonging to any schema over time. Holland and other researchers [8,9,35] proved an important result for GAs, known as the *schema theorem*. One possible simple interpretation for this result is the following: if the average fitness of individuals matching a schema H is high, then there is a high probability that the number of individuals matching H in the population at generation $t+1$ is larger than the number of individuals matching H at generation t. In other words, good schemata tend to multiply in the population along with generations, and thus are combined into good overall solutions with other such schemata. More recently, Stephens and co-workers [29,30] have produced *exact* formulations of the schema theorem, i.e., formulations which allow us to exactly predict the number of individuals sampling a given schema in a population at a certain generation, thus allowing us to draw important conclusions about the ability of the GA system to find satisfactory solutions.

A formalization of a schema theorem for GP, similar to Holland's theorem for GAs, is due to O'Reilly [14–16]. Other interesting schema theorem formulations for GP are due to Whigham [32–34], Rosca [24], and Poli and Langdon [18,19]. The development of an exact and general schema theory for GP is due to Poli and co-workers [13, 20–22], and it undoubtedly represents the most significant result in GP theory.

2.3 Conclusions

Given the number of tools that evolutionary computation offers researchers, engineers, and practitioners, each of which has its own peculiarities, strengths, and weaknesses, EC research is still very busy understanding the subtleties which are often hidden behind EC's apparently simple and theoretically well-defined approaches. Unfortunately, nothing comes for free, and this applies also to EC, which, of course, does not provide (yet?) fully automated and autonomous problem solvers. EC techniques do relieve designers from the burden of searching huge and rough spaces blindly for well-hidden solutions, and have few requirements in terms of domain-specific knowledge. However, at the same time, they also introduce users to new, typically smaller, but by no means less perilous domains in which, for instance, they are required to find the right parameters to let evolutionary algorithms perform at their best. This requires another, more general, kind of knowledge about the algorithms themselves. Too often EC researchers have to listen to the complaints of practitioners who, after applying EC techniques blindly and with overexpectations about something that can 'magically' solve their problems, declare that they just 'do not work'.

Every activity needs its specific tools, and every tool requires the acquisition of specific skills, possibly even easy to learn, before it can be used properly and effectively. Moreover, such skills can be taught theoretically, but some good examples by more experienced users and some direct practice are essential for learning. Within this book, this chapter has tried to provide the reader with some general knowledge about EC to start with; the next chapter will offer hints about what EC techniques are best suited to which applications in the medical domain, and the following chapters will provide some very nice examples of what can be done with a knowledgeable use of EC.

That is what our book can offer. We hope that its reading will be enough to arouse the curiosity of researchers and practitioners in the medical field, as well as inducing at least some of them to explore the EC world and take the chances it offers them. Finally, and most of all, we do hope that their exploration will be more effective and less deceptive after reading this.

References

[1] C. Darwin. *On the Origin of Species by Means of Natural Selection*. John Murray, 1859.

[2] K. A. De Jong. Genetic algorithms: a 10 year perspective. In J. J. Grefenstette (ed.), *Proceedings of an International Conference on Genetic Algorithms and Their Applications*. Erlbaum, 1985.

[3] K. A. De Jong. On using genetic algorithms to search program spaces. In J. J. Grefenstette (ed.), *Genetic Algorithms and Their Applications: Proceedings of the Second International Conference on Genetic Algorithms*. Erlbaum, 1987.

[4] K. A. De Jong. Learning with genetic algorithms: an overview. *Machine Learning*, 3:121–138, 1988.

[5] M. Dorigo, M. Birattari, and T. Stützle. Ant colony optimization – artificial ants as a computational intelligence technique. *IEEE Computational Intelligence Magazine*, 1(4):28–39, 2006.

[6] M. Dorigo, V. Maniezzo, and A. Colorni. Optimization by a colony of cooperating agents. *IEEE Transactions on Systems, Man, and Cybernetics – Part B*, 26(1):29–41, 1996.

[7] L. J. Fogel, A. J. Owens, and M. J. Walsh. *Artificial Intelligence through Simulated Evolution*. John Wiley & Sons, 1966.

[8] D. E. Goldberg. *Genetic Algorithms in Search, Optimization and Machine Learning*. Addison-Wesley, 1989.

[9] J. H. Holland. *Adaptation in Natural and Artificial Systems*. The University of Michigan Press, 1975.

[10] J. H. Holland, L. B. Booker, M. Colombetti, M. Dorigo, D. E. Goldberg, S. Forrest *et al*. What is a Learning Classifier System? In P. L. Lanzi, W. Stolzmann, and S. W. Wilson (eds), *Learning Classier Systems. From Foundations to Applications*, Vol. 1813 in LNAI, pp. 3–32. Springer-Verlag, 2000.

[11] J. Kennedy and R. Eberhart. Particle swarm optimization. In *Proceedings of the IEEE International Conference on Neural Networks*, Vol. IV, pp. 1942–1948. IEEE Computer Society Press, 1995.

[12] J. R. Koza. *Genetic Programming*. The MIT Press, 1992.

[13] W. B. Langdon and R. Poli. *Foundations of Genetic Programming*. Springer, 2002.

[14] U.-M. O'Reilly. An Analysis of Genetic Programming. PhD thesis, Carleton University, Ottawa, Canada, 1995.

[15] U.-M. O'Reilly and F. Oppacher. Using building block functions to investigate a building block hypothesis for genetic programming. Technical Report 94-02-029, Santa Fe Institute, Santa Fe, NM, 1994.

[16] U.-M. O'Reilly and F. Oppacher. The troubling aspects of a building block hypothesis for genetic programming. In L. D. Whitley and M. D. Vose (eds), *Foundations of Genetic Algorithms*, Vol. 3, pp. 73–88. Morgan Kaufmann, 1995.

[17] R. Poli, J. Kennedy, and T. Blackwell. Particle swarm optimization. *Swarm Intelligence*, 1(1):33–57, 2007.

[18] R. Poli and W. B. Langdon. A new schema theory for genetic programming with one-point crossover and point mutation. In J. R. Koza, K. Deb, M. Dorigo, D. B. Fogel, M. Garzon, H. Iba, and R. L. Riolo (eds), *Genetic Programming 1997: Proceedings of the Second Annual Conference*, pp. 278–285. Morgan Kaufmann, 1997.

[19] R. Poli and W. B. Langdon. Schema theory for genetic programming with one-point crossover and point mutation. *Evolutionary Computation*, 6(3):231–252, 1998.

[20] R. Poli and N. F. McPhee. Exact schema theorems for GP with one-point and standard crossover operating on linear structures and their application to the study

of the evolution of size. In J. Miller, M. Tomassini, P. L. Lanzi, C. Ryan, A. Tetta-manzi, and W. Langdon (eds), *Genetic Programming, Proceedings of EuroGP'2001*, Vol. 2038 of LNCS, pp. 126–142. Springer-Verlag, 2001.

[21] R. Poli and N. F. McPhee. General schema theory for genetic programming with subtree swapping crossover: Part I. *Evolutionary Computation*, 11(1):53–66, 2003.

[22] R. Poli and N. F. McPhee. General schema theory for genetic programming with subtree swapping crossover: Part II. *Evolutionary Computation*, 11(2):169–206, 2003.

[23] I. Rechenberg. *Evolutionsstrategie: Optimierung technischer Systeme nach Prinzipien der biologischen Evolution*. Fromman-Holzboog Verlag, 1973.

[24] J. P. Rosca. Analysis of complexity drift in genetic programming. In J. R. Koza, K. Deb, M. Dorigo, D. B. Fogel, M. Garzon, H. Iba, and R. L. Riolo (eds), *Genetic Programming 1997: Proceedings of the Second Annual Conference*, pp. 286–294. Morgan Kaufmann, 1997.

[25] C. Ryan, J. J. Collins, and M. O'Neill. Grammatical evolution: Evolving programs for an arbitrary language. In W. Banzhaf, R. Poli, M. Schoenauer, and T. Foga-rty (eds), *Genetic Programming, First European Workshop, EuroGP'98*, Vol. 1391 in LNCS, pp. 83–96. Springer, 1998.

[26] H.-P. Schwefel. Evolutionsstrategie und numerische optimierung, 1975. Dr.-Ing. Diss. Technical University of Berlin, Department of Process Engineering, Berlin.

[27] H.-P. Schwefel. Numerische optimierung von computer-modellen mittels der evolutionstrategie. *Interdisciplinary Systems Research*, Vol. 26. Birkhauser, 1977.

[28] H.-P. Schwefel. *Numerical Optimization of Computer Models*. John Wiley & Sons, 1981.

[29] C. R. Stephens and H. Waelbroeck. Effective degrees of freedom in genetic algorithms and the block hypothesis. In Back, T. *et al.* (eds), *Genetic Algorithms: Proceedings of the Seventh International Conference*, pp. 34–40. Morgan Kaufmann, 1997.

[30] C. R. Stephens and H. Waelbroeck. Schemata evolution and building blocks. *Evolutionary Computation*, 7(2):109–124, 1999.

[31] R. J. Urbanowicz and J. H. Moore. Learning classifier systems: A complete introduction, review, and roadmap. *Journal of Artificial Evolution and Applications*, 2009: 1–25, 2009.

[32] P. A. Whigham. Grammatical Bias for Evolutionary Learning. PhD thesis, School of Computer Science, University College, University of New South Wales, Australian Defence Force Academy, Canberra, Australia, 14 October 1996.

[33] P. A. Whigham. A schema theorem for context free grammars. In *Proceedings of the IEEE Conference on Evolutionary Computation*, Vol. 1, pp. 178–181. IEEE Press, 1995.

[34] P. A. Whigham. Search bias, language bias, and genetic programming. In J. R. Koza, D. E. Goldberg, D. B. Fogel, and R. L. Riolo (eds), *Genetic Programming 1996: Proceedings of the First Annual Conference*, pp. 230–237. MIT Press, 1996.

[35] D. Whitley. A genetic algorithm tutorial. *Statistics and Computing*, 4:65–85, 1994.

3

A Review of Medical Applications of Genetic and Evolutionary Computation

Stephen L. Smith

Department of Electronics, University of York, York, UK

In this chapter an overview is given of recent publications detailing the application of GEC to medicine. To provide some structure, papers have been classified into five main sections: Medical Imaging and Signal Processing; Data Mining Medical Data and Patient Records; Clinical Expert Systems and Knowledge-based Systems; Modelling and Simulation of Medical Processes; and Clinical Diagnosis and Therapy. It should be stressed that this classification is primarily for the convenience of the reader and, inevitably, there are going to be situations where papers could easily be classified into more than one section.

Genetic and Evolutionary Computation: Medical Applications   Edited by Stephen L. Smith and Stefano Cagnoni
© 2011 John Wiley & Sons, Ltd

3.1 Medical Imaging and Signal Processing

3.1.1 Overview

Medical imaging is widely accepted as a special case of signal processing dealing with two- (or three-dimensional) data in the form of conventional images, such as the common X-ray, but can also encompass other forms of medical signals not necessarily two-dimensional, an example of which may be the electrocardiogram (ECG), in which the data may be interpreted as a graph, map or image.

The application of computers to medical imaging is as longstanding as the imaging modalities themselves, and usually has one of two broad aims: either to process the image to allow better interpretation by a human observer, such as a health professional for the purpose of aiding diagnosis; or to pre-process the image as part of a machine-based diagnostic system. The processes involved in achieving these aims are not necessarily exclusive, but the results are usually substantially different.

The wide range of imaging modalities commonly used in clinical practice include classical X-ray-based radiology; computed tomography (CT), which requires the three-dimensional reconstruction of images from a set of two-dimensional radiographs; magnetic resonance imaging (MRI), a three-dimensional imaging modality more sensitive to soft tissues; positron emission tomography (PET) and single photon emission computed tomography (SPECT); and ultrasound. Their use in clinical practice and medical research has led to a plethora of imaging applications, which is both exciting and challenging. The full breadth of traditional techniques has been applied, including statistical, morphological, structural and grammatical image processing, as well as processing in the spatial and frequency domains. More recently, techniques such as fuzzy logic and neural networks have also been applied to medical imaging.

This section provides an overview of recent publications on genetic and evolutionary computation applied to medical imaging and in particular, image segmentation, registration, reconstruction and correction; other applications are also considered briefly.

3.1.2 Image Segmentation

The aim of image segmentation is to partition the image into distinct regions or objects for the purpose of subsequent interpretation, whether by the human observer or further machine processing, leading to a diagnosis.

Segmentation of medical images to identify anatomical structures of interest is particularly challenging due to the nature of the images and the structures themselves. A number of different approaches are traditionally employed, including histogram-based methods, clustering, edge detection, deformable models, region growing, level set and graph partitioning.

Previous work on applying genetic and evolutionary computation to image segmentation of medical images can be divided broadly into two categories: those that employ the evolutionary algorithms as an optimiser within a traditional segmentation method and those that present a novel process wholly or partly based around the evolutionary algorithm.

An example of work in which the algorithm is used as an optimiser is reported by Huang and Bai [1], where an improved genetic algorithm (GA) is used to optimise the two-dimensional Otsu algorithm [2] which automatically performs histogram shape-based thresholding. The enhancements to the GA concern the crossover operator which is applied to an individual based on its ranking in relation to the mean fitness, rather than fitness alone, and a dual mutation operator, which selects two individuals and applies separate logical operations to each. Applied to CT images of the ankle, the authors claim improved segmentation results and computational efficiency over the original Otsu algorithm. Another use of the GA as an optimiser is reported by Ballerini and Bocchi [3] in the automated segmentation of radiograms (X-rays) for the assessment of skeletal age, currently a difficult and time-consuming process. The GA optimises planar active contours, referred to as *snakes*, for each bone in the image. Neighbouring contours are coupled by an elastic force representing the physical relationship between anatomical regions. Segmented images are compared with hand-drawn outlines and provide encouraging results. An example of segmentation results for the three phalanx bones of the finger is shown in Figure 3.1.

In the segmentation of MRI brain images using genetically guided clustering by Sasikala, Kumaravel and Ravikumar [4], the GA optimises the objective function of the standard fuzzy c-means (FCM) algorithm to compensate for spatial intensity inhomogeneities, commonly associated with MRI. The results are shown in Figure 3.2.

Segmentation of the brain's ventricular system is important in the diagnosis of medical abnormalities. Levman, Alirezaie and Khan [5] implement a multi-species GA to identify the ventricles which comprises three structures: the first and second are for the detection of vertical edges and horizontal edges, respectively, and the third is for local information suppression of edge map information generated by the first two. The algorithm was applied to MTRI images from 14 patients and the authors report promising results.

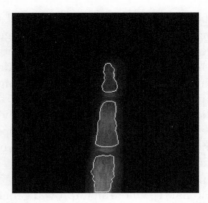

Figure 3.1 Segmentation results for the three phalanx bones of the finger (With kind permission from Springer Science+Business Media: *Lecture Notes in Computer Science*, "Multiple Genetic Snakes for Bone Segmentation," **2611**, © 2003, 346–356, Lucia Ballerini and Leonardo Bocchi)

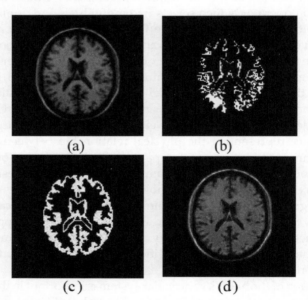

Figure 3.2 Comparison of segmentation results on MR image (a), original image corrupted with 40% intensity inhomogeneity and 9% Gaussian noise (b), segmented grey matter using a genetically guided fuzzy c-means algorithm (c) and segmented grey matter and bias field estimate using genetically guided bias-corrected fuzzy c-means algorithm GGBCFCM (d) (Reproduced with permission from Sasikala M, Kumaravel N, Ravikumar S., "Segmentation of brain MR images using genetically guided clustering," *Conf Proc IEEE Eng Med Biol Soc.* **1**:3620–3. © 2006 IEEE)

Two papers are now considered in which the evolutionary algorithms are used in conjunction with deformable shape models to achieve segmentation of the images. McIntosh and Hamarneh [6] use a GA to alleviate weaknesses in the initialisation, parameter selection and local minima by the evolution of a large number of models. The size and shape of the deformable models used are driven in terms of a learned shape which reduces the search space to be addressed, thus improving computational efficiency as demonstrated on mid-sagittal MRI scans. A full review of their work is given in Chapter 4.1 of this book. Heimann *et al.* [7] employ a real-valued vector-based evolutionary algorithm to initialise the shape-guided deformable model. The algorithm also differs from a standard GA in that no crossover operator is used and mutation is achieved by adding a random vector from a multivariate zero-mean Gaussian distribution. The technique was applied to 54 CT images of the liver and generated reliable segmentation of the organ in close agreement with manual tracings.

Two further non-standard approaches considered here are hierarchical evolutionary algorithms and artificial ant colonies. Lai and Chang [8] propose the hierarchical evolutionary algorithm for the segmentation of a range of CT and MRI images, including the brain and abdomen, and claim benefits over Hopfield neural networks, dynamic thresholding, k-means and fuzzy c-means algorithms. The hierarchical evolutionary algorithm is similar to a GA but contains control genes as well as parametric genes in a hierarchical structure. The control gene simply determines which parametric gene should be enabled and which should be disabled in the evolution process. Huang, Cao and Luo [9] propose the use of an artificial ant colony in the segmentation of MRI brain scans. The evolutionary algorithm is inspired by food search behaviour of ants and the associated pheromone deposition fields. The ants behave according to a simple rule set and search for food defined as structural objects within the image. After a number of iterations the deposited pheromone fields are used to segment the image as illustrated in Figure 3.3. Finally, Ma, Zhang and Hu [10] describe the use of a GA for edge patching following Canny edge detection and labelling, in the segmentation of the glomerulus, the filtering unit of the kidney.

3.1.3 Image Registration, Reconstruction and Correction

There are a number of applications in medical imaging where two or more images need to be aligned or registered to enable operations such as subtraction to be undertaken. Examples of common applications are to measure tissue mass changes over a period of time in the monitoring of cancerous

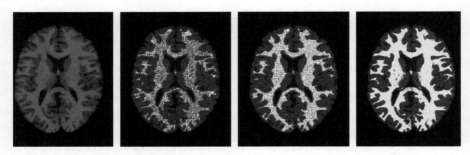

Figure 3.3 Segmentation results: (from left to right) the original MR image; the segmentation results using the AC algorithm for different iterations (Reprinted from "An artificial ant colonies approach to medical image segmentation," **92**, P. Huang, H. Cao, S. Luo, *Computer Methods and Programs in Biomedicine*, 267–273 , with permission from Elsevier)

growths and the presence of a radioactive tracer indicating vascular blood flow in angiography. Traditional approaches can be categorised as being either area based or landmark based and employing areas of image similarity or dissimilarity, respectively, to align the images.

Zhang *et al.* [11] employ a modified GA in conjunction with the concept of *mutual information* taken from information theory. The assumption is that the mutual information should be at a maximum when the two images are perfectly aligned. There are many traditional approaches to optimisation of mutual information matching schemes, but these suffer from local maxima. To address this problem, the modified GA employs adaptive crossover and mutation schemes which are applied to varying numbers of individuals. A multi-resolution optimisation strategy is also employed to reduce the search space and hence the computational effort required. MRI, CT and PET images are used in the evaluation of the technique, which the authors report to be both accurate and efficient.

Mañana, González and Romero [12] use a distributed adaptive GA to find an optimal affine transformation for intra-oral radiographs. The GA's genome has four floating point numbers representing the parameters used in the affine transformation, namely the scale factor, rotational angle and translations in the X and Y planes. The crossover operator is a variation of the averaging crossover operator and mutation, real number creep, adding or subtracting Gaussian distributed random noise to individuals. The results of their approach are shown in Figure 3.4.

Local image registration, as the name suggests, concentrates on localised areas of the image to perform the registration. This is particularly appropriate

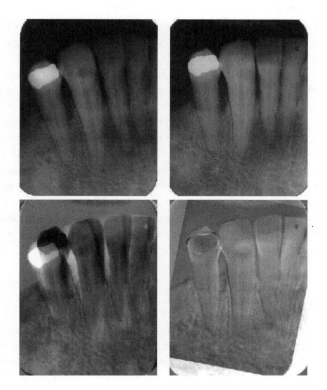

Figure 3.4 The upper row shows the two images to subtract. The bottom row shows the subtracted images: left without geometrical correction and right after automatic correction (Reproduced with permission from Manana, G.; Romero, E.; Gonzalez, F., "A Grid Computing Approach to Subtraction Radiography," *Image Processing*, 2006 IEEE International Conference on, 8–11 Oct. 2006, 3325–3328. © 2006 IEEE)

for some medical images which only differ in small areas. Peng *et al.* [13] determine a deformation function based on a radial basis function with compact support using a niche genetic algorithm. This GA uses decimal coding to represent the parameters of the deformation function, and a niche selection strategy that selects individuals based on the Euclidean distance between them. Standard crossover and mutation operators are also employed.

A multi-objective optimisation strategy implemented on hardware (using a field programmable gate array) for registration of medical images is reported by Dandekar *et al.* [14]. The implementation of a multi-objective algorithm presents a range of solutions for conflicting objectives including resources efficiency and accuracy. The performance benefits of this novel hardware implementation are a significant advance on similar software implementations.

The final two papers considered here specifically address the reconstruction and correction of medical images. Reconstruction of 3D images obtained from SPECT and PET scanners suffer from artefacts such as Compton scattering not experienced with other modalities. Bousquet, Louchet and Rocchisani [15] have developed an evolution strategy based on the Fly algorithm, originally used in real-time stereo sequencing processing, for a fully 3D reconstruction method that exploits all projection images. For the correction of cortical surfaces in MR images, Ségonne, Grimson and Fischl [16] have developed a GA with a representation that is particularly suited to the retessellation of the image; this GA also selects an initial population that will lead to early convergence and employs mutation and crossover operators that match the nature of the problem at hand.

3.1.4 Other Applications

Breast cancer is one of the most common cancers in women and a leading cause of death. Detection of breast cancers is traditionally made from mammograms (high-resolution X-rays of the breast), MRI or ultrasound images. However, Yao, Pan and Tao [17] report highly novel work on the application of a quantum-inspired GA applied to microwave imaging of the breast. The quantum GA combines the theory of quantum computing with evolutionary computing demonstrating strong parallelism and claims rapid convergence and high computational efficiency. Diagnosis of breast cancer based on classification of microcalcifications in conventional mammograms is considered by Hernandez-Cisneros, Terashima-Marin and Conant-Pablos [18], who use a GA for the selection of features which are then used as inputs to a feedforward neural network. The authors report improvements in overall accuracy, sensitivity and specificity of the classification over conventional class separability and forward sequential search.

Image compression is an important subject in medical imaging, where image resolutions and consequently storage requirements are high. One of the factors affecting compression performance of JPEG images is the quantisation table. Wu [19] reports the use of a GA to search for a quantisation table design that contributes to better compression efficiency in terms of bit rate and decoded quality. The results claim the GA-based search procedures generate better performance than JPEG 2000 and JPEG.

3.2 Data Mining Medical Data and Patient Records

With the widespread adoption of computer-based health systems, the digitisation of medical data, images and patient notes has increased significantly. New techniques are required to search and explore this data in a more effective and efficient manner, to identify relationships and patterns that can be used to aid diagnosis and treatment of patients. Data mining is the process of obtaining useful information from identifying these patterns and relationships, and is commonly associated with the following types of operation: classification – allocating data to a predefined class or category; clustering – the allocation of data into groups that have not been predefined; associations and rule learning – the formalisation of relationships between data. A range of techniques are traditionally used to mine data, including neural networks, decision trees, clustering algorithms and rule induction. GEC has been used in combination with these and as a stand alone method.

Chiu *et al.* [20] report the use of a GA with an association rule algorithm to mine three-dimensional anthropometric (body) scanning data and other medical profiles of patients to detect hypertension. The hybrid algorithm is used to discover classification trees to make this association and the approach is shown to outperform the use of a standard GA alone.

Ghannad-Rezaie *et al.* [21] propose a new medical data mining approach that integrates a classifier with a particle swarm algorithm for surgery candidate selection in temporal lobe epilepsy. Figure 3.5 summarises the results obtained for the particle swarm optimisation (PSO), compared with a range of methods: ant colony optimisation (ACO), Bayesian Belief Network (BBN), C4.5 (a decision tree approach) and LOcal LInear MOdel Tree (LOLIMOT). The area under the receiver operating characteristic (ROC) curve for each method relates to the accuracy of its classification performance.

Yin, Cheng and Zhou [22] employ a metaheuristic approach to data mining of medical datasets using the shuffled frog leaping (SFL) algorithm, which combines genetic-based and social behaviour-based strategies. A number of individuals (frogs) are generated and evolved in groups or memeplexes which are combined to give a population of solutions that the authors claim provide superior classification performance.

The analysis of phrase patterns in radiology mammography reports has been used by Patton *et al.* [23,24] to find abnormal instances without the need of a training set. A GA was used to implement a maximum variation sampling technique exploiting the premise that abnormal reports should include unusual or rare words, thus making them dissimilar to normal reports. A full description of the authors' work is presented in Chapter 5.1.

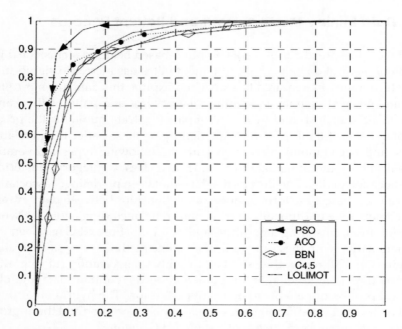

Figure 3.5 Results obtained for the particle swarm optimisation (PSO) using a receiver operating characteristic (ROC) curve, compared with a range of methods: ant colony optimisation (ACO), Bayesian Belief Network (BBN), C4.5 (a decision tree approach) and LOcal LInear MOdel Tree (LOLIMOT) (Reproduced with permission from Ghannad-Rezaie, M.; Soltanain-Zadeh, H.; Siadat, M.-R.; Elisevich, K.V., "Medical Data Mining using Particle Swarm Optimization for Temporal Lobe Epilepsy," *Evolutionary Computation*, 2006. CEC 2006. IEEE Congress on, 761–768. © 2006 IEEE)

Rajavarman and Rajagopalan [25] describe a two-phase data mining algorithm for use on a database of genetic features and environmental factors which are involved in multifactorial diseases. The first phase is a feature selection operation using a GA and the second, a k-means clustering algorithm. The GA also employs some advanced features: sharing, a niching mechanism which encourages individuals that occupy less crowded areas of the feature space and a random immigrant facility, where underperforming individuals within the population are replaced with a randomly generated individual to help maintain diversity.

The evolution of rules through a multi-objective evolutionary algorithm is proposed by Zaharie, Lungeanu and Zamfirache [26], which also permits the involvement of a user in the evolution process. This work is described fully in Chapter 5.2.

Tapaswi and Joshi [27] describe an evolutionary assisted computing method for the mining of biomedical images as a screening technique and in support of clinical diagnosis. The technique uses a combined vector of colour and texture features which is fed into a neural network, trained for classification using a GA. It is envisaged that subsequent clinical images can be screened for any suspected disease for which the algorithm has been trained.

Another application that uses an evolutionary algorithm to train neural networks to sample large, unbalanced datasets is reported by Lu *et al.* [28]. A novel approach to selective sampling derived from the estimation-exploration algorithm is applied to an unbalanced dataset of over 1 million records detailing characteristics of trauma patients and their subsequent condition. The results obtained suggest the approach outperforms random sampling and balanced sampling that are widely used for datasets of this type.

A visual mining data paradigm employing genetic programming and classical optimisation is proposed by Valdés, Orchard and Barton [29] and is considered in detail in Chapter 5.3.

Winkler, Affenzeller and Wagner [30] report the development of a genetic programming technique combining logical expressions and mathematical functions for evolving classifiers that have been evaluated on five medical datasets including heart, diabetes, thyroid and melanoma data.

3.3 Clinical Expert Systems and Knowledge-based Systems

Expert and knowledge-based systems often employ rule-based, case-based and model-based reasoning to extract knowledge from raw data. Here, work is reviewed which applies GEC, often in combination with other computational techniques, to medical planning, diagnosis and treatment.

A comprehensive and recent review of knowledge-based and intelligent computing systems applied to medicine is provided by Pandey and Mishra [31], and encompasses a range of techniques such as model-based reasoning, artificial neural networks, fuzzy systems as well as GAs.

Krajnak and Xue [32] employ a GA optimised fuzzy system to report patient status monitoring in the operating theatre based on metrics such as change in blood pressure and heart rate. Specifically, GA optimises the fuzzy rule weights, outputs and input membership and its fitness function based on the area under the receiver operating characteristic (ROC) curve.

Grosan *et al.* [33] describe the application of a simple evolutionary algorithm to the multi-criterion problem for which a conventional Pareto dominance relationship between solutions is not appropriate, in this case for ranking treatment methods for a condition that can cause excruciating pain to the face and head, known as trigeminal neuralgia. The EA simply evolves a hierarchy of treatments where each chromosome comprises strings representing combinations of treatments used. This is implemented using simple selection and mutation operators and a fitness function based on treatment efficiency.

Koutsojannis and Hatzilygeroudis [34] describe the use of a fuzzy expert system that incorporates an evolutionary algorithm for the diagnosis and treatment of blood gas disturbances and disorders observed in intensive care unit patients. Diagnosis of these disturbances and disorders of the patient's blood traditionally rely upon both laboratory data and clinical observations. The fuzzy expert system uses a differential evolutionary algorithm, which creates new candidates based on the difference between two population vectors, rather than a standard mutation operator.

Sprogar, Sprogar and Colnari [35] describe the development of an evolutionary algorithm that combines ideas from EC, systems theory and the theory of chaos for constructing decision trees. The algorithm requires no or minimal human interaction and shows some interesting properties when used on different medical datasets. The algorithm uses a non-standard implicit fitness evaluation and the capability to self-adapt to a given problem.

The identification of specific medical cases in addition to, or in place of, a generic model is often made through clinical findings such as risk factors, signs and symptoms, as well as the final diagnosis. Kléma, Kubalík and Lhotská [36] use a genetic algorithm in the implementation of an instance-based reasoning paradigm for mortality prediction of patients following cardiological intervention.

Nuovo and Catania [37] optimise a fuzzy clustering algorithm with a GA to generate a rule-based system which can be used for medical diagnosis. Datasets for three medical conditions – breast cancer, diabetes and aphasia – are used to test the effectiveness of the technique, both in terms of its accuracy and its easy-to-read rules.

Paul et al. [38] consider the application of evolutionary algorithms for the identification of health risk factors aimed at preventing metabolic syndrome, a combination of medical disorders that increases the risk of developing cardiovascular disease and diabetes. Two evolutionary algorithms are compared, a standard GA and a variant called a random probabilistic model building genetic algorithm (RPMBGA). RPMBGA is a global search heuristic which is similar to a GA but maintains a vector of probabilities of the features being optimised and uses these to generate new individuals instead of using traditional crossover and mutation operators.

The prediction of coronary heart disease complications in type 2 diabetes mellitus patients is reported by Giardina *et al.* [39]. They describe the use of a GA and weighted k-nearest neighbours algorithm to identify potential risk factors from physical and laboratory examination results obtained from patient consultations. Although only modest improvements in using this approach over more conventional techniques were reported, this may be due to factors associated with the integrity of the data and the treatment the patients were receiving.

Another paper investigating myocardial infarction as a side-effect of medication for diabetic patients is reported by Sordo, Ochoa and Murphy [40]. They employ a particle swarm optimisation/ant colony optimisation methodology on a set of 13 risk factors comprising demographic and medical data obtained from electronic patient records. Not only were the results consistent with those reported in the medical literature, but the technique also lends itself to a graphical representation that is easy to understand and automatically generated by the system.

3.4 Modelling and Simulation of Medical Processes

Modelling and simulation are of great importance in a wide range of medicine and healthcare areas, including training, research and optimisation of procedures. GEC, often in combination with other techniques such as support vector machines and fuzzy-based learning, is used to great effect. Applications are particularly widespread and range from predicting the incidence of disease to modelling of foetal electrocardiograms, optimising the throughput of patients in a hospital emergency department to modelling visuo-spatial ability in Alzheimer's disease patients.

Fu *et al.* [41] use a support vector machine (SVM) and GA to determine the relationship between climate factors and incidence of dengue fever, a life-threatening disease that occurs widely in the tropics and southern China. The GA-based SVM is able to deal with multiple climatic factors and the time lag that is experienced in the development of dengue fever. The authors find that although all climatic features may influence the process, temperature and rainfall are particularly indicated, in line with other published research. The results are summarised in Figure 3.6.

Visceral leishmaniasis is another deadly parasitic disease that forms granuloma structures in the liver. Flugge *et al.* [42] have used an artificial immune system to model this behaviour with UML, a language that uses graphical notation, with agents. Initial results are encouraging and support the

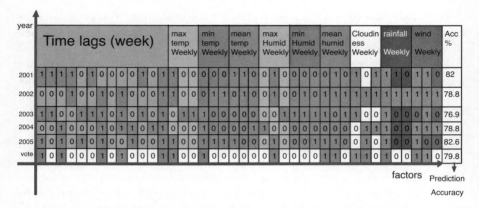

Figure 3.6 GA-based SVM classification model summarising the influence of climate factors on the dengue fever incidence trends. A '1' associated with a particular climate factor indicates influence on the time lag of dengue fever in the respective week and year (Reproduced with permission from Xiuju Fu; Liew, C.; Soh, H.; Lee, G.; Hung, T.; Lee-Ching Ng, "Time-series infectious disease data analysis using SVM and genetic algorithm," Evolutionary Computation, 2007. CEC 2007. 1276–1280 © 2007 IEEE)

development of such immunological models that are difficult to perform in vivo over periods of time.

A highly novel evolutionary hypernetwork is proposed by Ha *et al.* [43] to model cardiovascular disease and breast cancer using two real-valued datasets. The hypernetwork is a specific case of a hypergraph with weights assigned to the hyperedges. Simple learning and evolutionary computing mechanisms are then applied to generate a model with good classification characteristics with the real-valued datasets.

Atrial fibrillation is a common heart condition that can lead to myocardial infarction or stroke. Electrophysiological models have been developed to investigate the underlying electrical mechanisms of this condition and help predict the effect of specific medication. One drawback is that the models adopted are generic and yet there are significant variances in the electrophysiology between individuals. Syed, Vigmond and Leon [44] investigate the use of a GA to automatically fit these generic models to specific cases and find that the results are a reliable representation.

Another cardiology application, the foetal electrocardiogram, is an important indicator of foetal stress, maturity and position as well as providing diagnostic information on cardiac defects and arrhythmias. Nazarpour, Ebadi and Sane [45] propose a mathematical model whose parameters are optimised

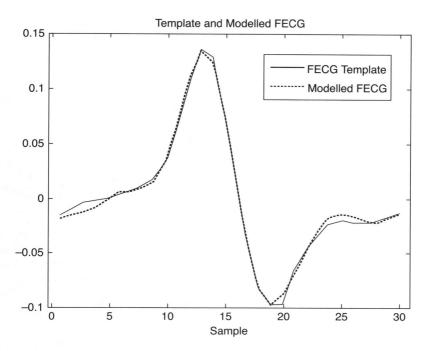

Figure 3.7 Modelled foetal electrocardiogram compared with an averaged (template) EC (Reproduced with permission from Nazarpour, K.; Ebadi, S.; Sane, S., "Fetal Electrocardiogram Signal Modelling Using Genetic Algorithm," Medical Measurement and Applications, 2007. MEMEA '07. © 2007 IEEE)

using a standard GA. The resulting curve employs nine coefficients and closely matches the training template, as shown in Figure 3.7.

Bosman and Alderliesten [46] describe the use of an evolutionary algorithm in the design of a medical simulation for minimally invasive vascular intervention, which is becoming commonly used to train clinicians. The simulation is based on an optimisation problem, the minimisation of total energy associated with the trajectory of the guide wire used to perform the invasive vascular intervention. An iterative density estimation evolutionary algorithm, which differs from conventional EAs in that probabilistic models are learned using the selected solutions, is used in conjunction with a local optimiser to speed up convergence. The result is an accurate and closer to real-time simulation of the intervention. An example depicting the modelling of an invasive guide wire as it is navigated towards the vascular abnormality is shown in Figure 3.8, illustrating the performance of the gradient leveraged iterated density evolutionary (GLIDE) algorithm developed by the authors.

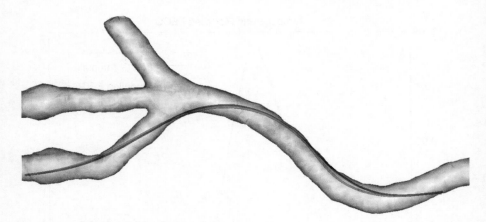

Figure 3.8 Modelling of an invasive guide wire as it is navigated towards the vascular abnormality; the reference guide wire configuration (black) and the simulated guide wire configuration (grey) using the gradient leveraged iterated density evolutionary (GLIDE) algorithm (Reproduced with permission from Alderliesten, T.; Bosman, P.A.N.; Niessen, W.J.; , "Towards a Real-Time Minimally-Invasive Vascular Intervention Simulation System," Medical Imaging, IEEE Transactions on, **26**, 1, 128–132, Jan. 2007 © 2007 IEEE)

Visuo-spatial ability, required to undertake many everyday tasks, including making simple three-dimensional drawings, is an important indicator of neurodegenerative conditions such as Alzheimer's disease. Smith and Lones [47] use a form of genetic programming called implicit context representation Cartesian genetic programming to model the maturing visuo-spatial ability in children between the ages of 7 and 11 years. This is achieved by digitising the children's drawings in real-time using a commercial digitising tablet and presenting the acceleration profile of the pen movements directly to the algorithm. It is argued that this model of maturing visuo-spatial ability can be used in reverse to predict the loss of visuo-spatial ability in Alzheimer's disease patients. A full description of this work can be found in Chapter 6.1.

Wu *et al.* [48] use a GA-based fuzzy learning system to model urinary analysis, an important and common clinical test used to diagnose diseases of the kidney and urinary tract. Specifically, the authors propose an algorithm to reduce the number of urine samples required for urine flow cytometry and conditions under which microscopic reviews are required to confirm certain clinical conditions. Their results suggest that the rules generated are easy to understand and provide a better classification accuracy than is currently achievable.

Head and neck squamous cell carcinoma is an oral cancer associated with smoking and alcohol consumption. However, the individual risk is also thought to be modified by the patient's genetic profile. Passaro *et al.* [49] use XCS, an evolution of Holland's learning classifier system, to generate a rule set to model this relationship. The ultimate aim is to identify those genes involved in susceptibility to oral cancer and highlight any interactions between them. The use of XCS was successful in adapting to data of different types and classification accuracy was higher than that obtained using a standard algorithm.

The modelling of biologically plausible behaviour in a neuromuscular simulation of a biceps/triceps pair is described using a GA and particle swarm optimiser by Gotshall *et al.* [50]. The results obtained are compared to those obtained from human subjects and demonstrate that the algorithms are effective in producing plausible models at both neural and gross anatomical levels. Another simulation-based study by Howard, Tyrrell and Cooper [51] uses an evolutionary algorithm to model the oral tract (mouth) shapes required to produce a set of vowel sounds. This work is reported fully in Chapter 6.2.

Accurate prediction of healthcare costs is becoming increasingly important. Stephens *et al.* [52] compare the use of a GA with conventional statistical measures to predict the future 0.5% patients that will incur the highest healthcare costs. The GA-based approach was shown to give significant performance gains compared with two standard industry benchmarks and also allows for non-linear interactions between attributes to be incorporated.

Petrovski and McCall [53] describe a problem-solving environment for modelling, simulation and optimisation of cancer chemotherapy. Both a GA and particle swarm optimisation algorithms are implemented in the distributed environment which employs web services to acquire the relevant data. In another study aimed at improving efficiency, Yeh and Lin [54] describe the development of a GA optimised simulation technique to reduce the waiting time for patients attending a hospital emergency department. The simulation considers the time spent by the patient through each relevant stage of the process from arrival, such as triage, emergency room, laboratory tests and bandaging station. The algorithm's solution is in the form of a revised rostering schedule for the nursing staff and reports a reduction in waiting time by approximately 43%.

Finally, Dumas and El Alaoui describe how genetic algorithms can be used to determine the optimal position of electrodes for a pacemaker using simulations that can avoid resource-intensive invasive techniques. A full description of their work is provided in Chapter 6.3.

3.5 Clinical Diagnosis and Therapy

This section considers applications of GEC leading to clinical diagnosis and therapy that have not been considered in previous sections of this chapter, such as medical imaging, data mining and expert systems.

Kim *et al.* [55] report the implementation of an evolutionary algorithm-based signal processing filter on a reconfigurable hardware chip as part of the system for automated heart disease diagnosis. The electrocardiogram (ECG) is processed by a dynamically evolved filter to remove noise in preparation for further processing and then transmission to the doctor's computer or PDA for final diagnosis. A parallel GA evolves FIR filters to determine optimal parameters for the particular signal properties. Their system is illustrated in Figure 3.9.

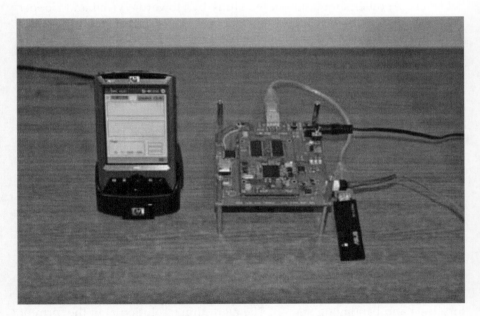

Figure 3.9 A system for automated heart disease diagnosis comprising an evolutionary algorithm-based signal processing filter implemented on reconfigurable hardware which then communicates results to the consultant's PDA (Reprinted from Tae Seon Kim. Hanho Lee. Jaehyun Park. Chong-Ho Lee. Yong-Min Lee. Chang-Seok Choi. Seung-Gon Hwang. Hyun Dong Kim. Chul Hong Min. "Ubiquitous evolvable hardware system for heart disease diagnosis applications," *International Journal of Electronics*, (http://www.informaworld.com), **95**, 7, 637–51. © 2008, with permission from Taylor & Francis Ltd)

In another application of evolutionary algorithms to ECGs, Goletsis *et al.* [56] use a GA to optimise thresholds and weight values of a multicriteria sorting method to classify the cardiac beats in the detection of myocardial ischaemic episodes.

Bhatia, Prakash and Pillai [57] describe a decision support system for the diagnosis and classification of heart disease. An integer-coded GA is employed to optimise the feature subset to maximise the support vector machine classification accuracy. The dataset used consists of 13 numeric attributes which include both physical and medical measurements. Results are encouraging and reported to better existing methods in some cases. In a similar study, Yan *et al.* [58] use a real-coded GA for the diagnosis and classification of five major heart diseases. The algorithm selects what is considered to be the 24 most critical features out of a set of 40 comprising patient symptoms and results from physical examination. A dataset of 352 cases is used to train the algorithm and the authors report that high accuracy results have been obtained.

The diagnosis of language impairments in speech pathology using a novel hybrid method combining competitive fuzzy cognitive maps with a GA is proposed by Stylios and Georgopoulos [59]. In an example application, 15 subjective attributes for speech and language pathology are used to diagnose specific language impairment, a disorder that can easily be confused with other language disorders.

The accurate diagnosis of Alzheimer's disease is of huge medical and social importance and yet no reliable test currently exists in clinical practice. Kim *et al.* [60] report the analysis of electroencephalograms (EEG) that map electrical activity on the scalp to diagnose Alzheimer's disease with the use of a GA and neural network. From the large number of quantitative features generated from the EEG, the combined GA/neural network is used to select the most effective subset of 35, which are then used as inputs to the neural network for subsequent classification.

Hazell and Smith [61] are developing a very different approach to the diagnosis of Alzheimer's disease, which will require patients to undertake a figure-copying task that is digitised in real-time. The data is fed into a special genetic program, a variant of a Cartesian genetic program, which is trained on a population of Alzheimer's patients and aged-matched controls. This paper uses drawings from a population of children aged between 7 and 11 years to simulate the immature performance expected of Alzheimer's patients. Results show the ability of the algorithm to discriminate between drawings by children of different ages. A similar method has been used by Smith *et al.* [62] to diagnose Parkinson's disease patients, but using alternative shapes for the patients to copy and different features to indicate the presence of distinguishing features. Figure 3.10 shows how a patient's drawing is digitised,

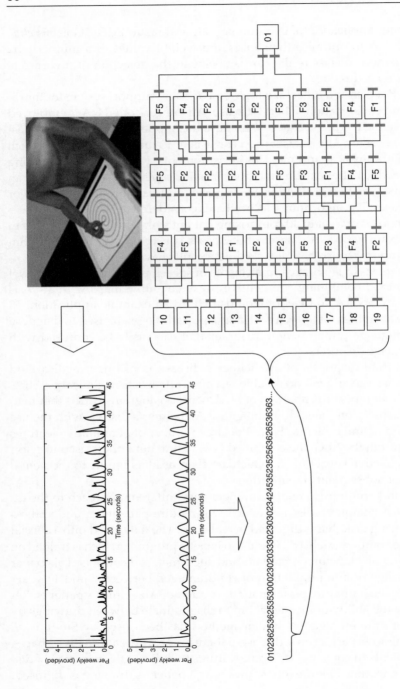

Figure 3.10 Digitisations of patients' drawings are pre-processed and input to a Cartesian genetic program which is trained to recognise the symptoms of Parkinson's disease

pre-processed and then fed into a Cartesian genetic program network which is trained to recognise the symptoms of Parkinson's disease.

The following papers provide examples in which GEC can be used to design, provide or optimise therapy for patients.

Determining optimum long-term dietary menu planning is an important therapeutic requirement for patients with a range of disorders. Gaál, Vassányi and Kozmann [63] propose an evolutionary divide-and-conquer method using a multi-objective GA for menu planning which has to satisfy a number of numerical as well as subjective criteria.

A therapeutic device called a stent is a permanent metallic implant used to open arteries blocked with atherosclerotic plaques. Blouza, Dumas and M'Baye [64] also use a multi-objective evolutionary algorithm to design a new stent that has optimal parameters to ensure blood flow through an artery that may avoid or at least reduce restenosis, the reoccurrence of the narrowing of the blood vessel.

Evolutionary algorithms are increasingly being used in drug design. Lameijer *et al.* [65] review the way in which evolutionary algorithms are used, both to create new molecules and to construct methods for predicting the properties of real or as yet non-existing molecules.

References

[1] Huang, T.-l. and Bai, X. 2008. An improved algorithm for medical image segmentation. WGEC '08. Second International Conference on Genetic and Evolutionary Computing, 25–26 September 2008, pp. 289–292. URL: http://ieeexplore. ieee.org/stamp/stamp.jsp?arnumber=4637447&isnumber=4637374.

[2] Otsu, N. 1979. A threshold selection method from gray-level histograms. *IEEE Trans. Syst., Man., Cyber.* 9:62–66. doi:10.1109/TSMC.1979.4310076.

[3] Ballerini, L. and Bocchi, L. 2003. Multiple genetic snakes for bone segmentation. *Lecture Notes in Computer Science*, Vol. 2611/2003, pp. 346–356.

[4] Sasikala, M., Kumaravel, N. and Ravikumar, S. 2006. Segmentation of brain MR images using genetically guided clustering. *Conf. Proc. IEEE Eng. Med. Biol. Soc.* 1:3620–3623.

[5] Levman, J., Alirezaie, J. and Khan, G. 2005. Magnetic resonance based ventricle system classification by multi-species genetic algorithm. Proc. 2nd Int. IEEE EMBS Conf. on Neural Engineering, 16–19 March 2005, pp. 2–5.

[6] McIntosh, C. and Hamarneh, G. 2006. Genetic algorithm driven statistically deformed models for medical image segmentation. GECCO '06, 8–12 July 2006, Seattle, WA.

[7] Heimann, T., Münzing, S., Meinzer, H.P. and Wolf, I. 2007. A shape-guided deformable model with evolutionary algorithm initialization for 3D soft tissue segmentation. *Inf. Process Med. Imaging* 20:1–12.

[8] Lai, C.-C. and Chang, C.-Y. 2009. A hierarchical evolutionary algorithm for automatic medical image segmentation. *Expert Syst. Appl.* **36**(1):248–259. doi:10.1016/j.eswa.2007.09.003.

[9] Huang, P., Cao, H. and Luo, S. 2008. An artificial ant colonies approach to medical image segmentation. *Comput. Meth. Progr. Biomed.* **92**(3):267–273.

[10] Ma, J., Zhang, J. and Hu, J. 2009. Glomerulus extraction by using genetic algorithm for edge patching. CEC '09. IEEE Congress on Evolutionary Computation, 18–21 May 2009, pp. 2474–2479.

[11] Zhang, H., Zhou, X., Sun, J. and Zhang, J. 2005. A novel medical image registration method based on mutual information and genetic algorithm. International Conference on Computer Graphics, Imaging and Vision: New Trends, 26–29 July 2005, pp. 221–226. URL: http://ieeexplore.ieee.org/stamp/stamp. jsp?arnumber=1521067&isnumber=32539.

[12] Mañana, G., González, F. and Romero, E. 2005. Distributed genetic algorithm for subtraction radiography. GECCO '05. Proceedings of the 2005 Workshops on Genetic and Evolutionary Computation, Washington, D.C., 25–26 June 2005. ACM, New York, pp. 140–146. doi:http://doi.acm.org/10.1145/1102256.1102288.

[13] Peng, W., Tong, R., Qian, G. and Dong, J. A local registration approach of medical images with niche genetic algorithm. CSCWD '06. 10th International Conference on Computer Supported Cooperative Work in Design, 3–5 May 2006, pp. 1–6. URL: http://ieeexplore.ieee.org/stamp/stamp.jsp?arnumber=4019194&isnumber=4019032.

[14] Dandekar, O., Plishker, W., Bhattacharyya, S. and Shekhar, R. 2008. Multiobjective optimization of FPGA-based medical image registration. FCCM. Proceedings of the 2008 16th International Symposium on Field-Programmable Custom Computing Machines, 14–15 April 2008. IEEE Computer Society, Washington, D.C., pp. 183–192. doi:http://dx.doi.org/10.1109/FCCM.2008.50.

[15] Bousquet, A., Louchet, J. and Rocchisani, J. 2008. Fully three-dimensional tomographic evolutionary reconstruction in nuclear medicine. *Lecture Notes in Computer Science*, Vol. 4926/2008, pp. 231–242.

[16] Ségonne, F., Grimson, E. and Fischl, B. 2005. A genetic algorithm for the topology correction of cortical surfaces. *Inf. Process Med. Imaging* **19**:393–405.

[17] Yao, M., Pan, Q. and Tao, Z. 2009. Application of quantum genetic algorithm on breast tumor imaging with microwave. GECCO '09. Proceedings of the 11th Annual Conference Companion on Genetic and Evolutionary Computation: Late Breaking Papers, Montreal, Canada, 8–12 July 2009. ACM, New York, pp. 2685–2688. doi:http://doi.acm.org/10.1145/1570256.1570383.

[18] Hernandez-Cisneros, R.R., Terashima-Marin, H. and Conant-Pablos, S.E. 2007. Comparison of class separability, forward sequential search and genetic algorithms for feature selection in the classification of individual and clustered microcalcifications in digital mammograms. ICIAR 2007, Proceedings of the 4th International Conference. *Lecture Notes in Computer Science*, Vol. 4633. Springer-Verlag, pp. 911–922.

[19] Wu, Y.-G. 2004. GA-based DCT quantisation table design procedure for medical images. Vision, Image and Signal Processing, IEE Proceedings, Vol. 151, no. 5, pp. 353–359, 30 October 2004. URL: http://ieeexplore.ieee.org/stamp/stamp.jsp?arnumber=1367349&isnumber=29931.

[20] Chiu, C., Hsu, K.-H., Hsu, P.-L., Hsu, C.-I., Lee, P.-C., Chiou, W.-K., Liu, T.-H., Chuang, Y.-C. and Hwang, C.-J. 2007. Mining three-dimensional anthropometric body surface scanning data for hypertension detection. *IEEE Trans. Inf. Technol. Biomed.* **11**(3):264–273.

[21] Ghannad-Rezaie, M., Soltanain-Zadeh, H., Siadat, M.-R. and Elisevich, K.V. 2006. Medical data mining using particle swarm optimization for temporal lobe epilepsy. IEEE Congress on Evolutionary Computation (CEC) 2006, pp. 761–768.

[22] Yin, H., Cheng, F. and Zhou, C. 2008. An efficient SFL-based classification rule mining algorithm. IEEE International Symposium on IT in Medicine and Education (ITME) 2008, 12–14 December, pp. 969–972.

[23] Patton, R.M., Beckerman, B. and Potok, T.E. 2008. Analysis of mammography reports using maximum variation sampling. GECCO '08. Proceedings of the 2008 GECCO Conference Companion on Genetic and Evolutionary Computation, Atlanta, GA, 12–16 July. ACM, New York, pp. 2061–2064. doi:http://doi.acm.org/10.1145/1388969.1389022.

[24] Patton, R.M., Potok, T.E., Beckerman, B.G. and Treadwell, J.N. 2009. A genetic algorithm for learning significant phrase patterns in radiology reports. GECCO '09. Proceedings of the 11th Annual Conference Companion on Genetic and Evolutionary Computation Conference: Medical Applications of Genetic and Evolutionary Computation (MedGEC), Montreal, Canada, 8–12 July. ACM, New York, pp. 2665–2670. doi:http://doi.acm.org/10.1145/1570256.1570380.

[25] Rajavarman, V.N. and Rajagopalan, S.P. 2007. Feature selection in data-mining for genetics using genetic algorithm. *J. Comput. Sci.* **3**(9):723–725.

[26] Zaharie, D., Lungeanu, D. and Zamfirache, F. 2008. Interactive search of rules in medical data using multiobjective evolutionary algorithms. GECCO '08. Proceedings of the 2008 GECCO Conference Companion on Genetic and Evolutionary Computation, Atlanta, GA, 12–16 July. ACM, New York, pp. 2065–2072. doi:http://doi.acm.org/10.1145/1388969.1389023.

[27] Tapaswi, S. and Joshi, R.C. 2004. An evolutionary computing approach for mining of biomedical images, LNCS 3333, pp. 523–532.

[28] Lu, Z., Rughani, A.I., Tranmer, B.I. and Bongard, J. 2008. Informative sampling for large unbalanced data sets. GECCO '08. Proceedings of the 2008 GECCO Conference Companion on Genetic and Evolutionary Computation, Atlanta, GA, 12–16 July. ACM, New York, pp. 2047–2054. doi:http://doi.acm.org/10.1145/1388969.1389020.

[29] Valdés, J.J., Orchard, R. and Barton, A.J. 2007. Exploring medical data using visual spaces with genetic programming and implicit functional mappings. GECCO '07. Proceedings of the 2007 GECCO Conference Companion on Genetic and Evolutionary Computation, London, UK, 7–11 July.

[30] Winkler, S.M., Affenzeller, M. and Wagner, S. 2009. Using enhanced genetic programming techniques for evolving classifiers in the context of medical diagnosis. *Genet. Progr. Evolv. Mach.* **10**(2):111–140. doi:http://dx.doi.org/10.1007/s10710-008-9076-8.

[31] Pandey, B. and Mishra, R.B. 2009. Knowledge and intelligent computing system in medicine. *Comput. Biol. Med.* **39**(3):215–230. doi:http://dx.doi.org/10.1016/j.compbiomed.2008.12.008.

[32] Krajnak, M. and Xue, J. 2006. Optimizing fuzzy clinical decision support rules using genetic algorithms. Engineering in Medicine and Biology Society (EMBS) '06. 28th Annual International Conference of the IEEE, 30 August–3 September, pp. 5173–5176.

[33] Grosan, C., Abraham, A., Tigan, S. and Chang, T-G. 2006. How to solve a multicriterion problem for which Pareto dominance relationship cannot be applied? A case study from medicine. Knowledge-Based Intelligent Information and Engineering Systems. 10th International Conference, KES 2006. Proceedings, Part III. *Lecture Notes in Artificial Intelligence*, Vol. 4253. Springer-Verlag, pp. 1128–1135.

[34] Koutsojannis, C. and Hatzilygeroudis, I. 2006. Fuzzy-evolutionary synergism in an intelligent medical diagnosis system. Knowledge-Based Intelligent Information and Engineering Systems. 10th International Conference, KES 2006. Proceedings, Part II. *Lecture Notes in Artificial Intelligence*, Vol. 4252. Springer-Verlag, pp. 1305–1312.

[35] Sprogar, M., Sprogar, M. and Colnari, M. 2005. Autonomous evolutionary algorithm in medical data analysis. *Comput. Meth. Progr. Biomed.* **80**:S29–S38.

[36] Kléma, J., Kubalík, J. and Lhotská, L. 2005. Optimized model tuning in medical systems. *Comput. Meth. Progr. Biomed.* **80**:S17–S28.

[37] Di Nuovo, A.G. and Catania, V. 2006. Genetic tuning of fuzzy rule deep structures for efficient knowledge extraction from medical data. SMC '06. IEEE International Conference on Systems, Man and Cybernetics 2006, Vol. 6, pp. 5053–5058, 8–11 October.

[38] Paul, T.K., Ueno, K., Iwata, K., Hayashi, T. and Honda, N. 2008. Genetic algorithm based methods for identification of health risk factors aimed at preventing metabolic syndrome. Proceedings of the 7th International Conference on Simulated Evolution and Learning, Melbourne, Australia, 7–10 December 2008. *Lecture Notes in Computer Science*, Vol. 5361. Springer-Verlag, pp. 210–219. doi:http://dx.doi.org/10.1007/978-3-540-89694-4_22.

[39] Giardina, M., Azuaje, F., McCullagh, P. and Harper, R. 2006. A supervised learning approach to predicting coronary heart disease complications in type 2 diabetes mellitus patients. Proceedings of the Sixth IEEE Symposium on Bioinformatics and Bioengineering (BIBE), 16–18 October 2006. IEEE Computer Society, Washington, DC, pp. 325–331.

[40] Sordo, M., Ochoa, G. and Murphy, S.N. 2009. A PSO/ACO approach to knowledge discovery in a pharmacovigilance context. Proceedings of the 11th Annual

Conference on Genetic and Evolutionary Computation: Medical applications of genetic and evolutionary computation (MedGEC). GECCO '09. ACM, New York, pp. 2679–2684.

[41] Fu, X., Liew, C., Soh, H., Lee, G., Hung, T. and Ng, L.-C. 2007. Time-series infectious disease data analysis using SVM and genetic algorithm. IEEE Congress on Evolutionary Computation (CEC) 2007, 25–28 September, pp. 1276–1280.

[42] Flugge, A.J., Timmis, J., Andrews, P., Moore, J. and Kaye, P. 2009. Modelling and simulation of granuloma formation in visceral leishmaniasis. IEEE Congress on Evolutionary Computation (CEC) 2009, 18–21 May, pp. 2123–2128.

[43] Ha, J.-W., Eom, J.-H., Kim, S.-C. and Zhang, B.-T. 2007. Evolutionary hypernetwork models for aptamer-based cardiovascular disease diagnosis. Proceedings of the 9th Annual Conference Companion on Genetic and Evolutionary Computation, London, UK. Workshop session: Medical Applications of Genetic and Evolutionary Computation (MedGEC), pp. 2709–2716.

[44] Syed, Z.F., Vigmond, E. and Leon, L.J. 2005. Suitability of genetic algorithm generated models to simulate atrial fibrillation and k+channel blockades. IEEE 27th Annual International Conference of the Engineering in Medicine and Biology Society (EMBS), 17–18 January 2005, pp. 7087–7090.

[45] Nazarpour, K., Ebadi, S. and Sane, S. 2007. Fetal electrocardiogram signal modelling using genetic algorithm. IEEE International Workshop on Medical Measurement and Applications (MEMEA) 2007, 4–5 May, pp. 1–4.

[46] Alderliesten, T., Bosman, P.A.N., Niessen, W.J. 2007. "Towards a Real-Time Minimally-Invasive Vascular Intervention Simulation System," Medical Imaging, IEEE Transactions on, vol. 26, no. 1, pp. 128–132, doi: 10.1109/TMI. 2006.886814 URL: http://ieeexplore.ieee.org/stamp/stamp.jsp?tp=&arnumber= 4039522&isnumber=4039521

[47] Smith, S.L. and Lones, M.A. 2009. Implicit context representation Cartesian genetic programming for the assessment of visuo-spatial ability. IEEE Congress on Evolutionary Computation, Trondheim, Norway, 2009.

[48] Wu, P., Goodman, E.D., Jiang, T. and Pei, M. 2009. A hybrid GA-based fuzzy classifying approach to urinary analysis modeling. Proceedings of the 11th Annual Conference Companion on Genetic and Evolutionary Computation, Montreal, Canada. Workshop session: Medical Applications of Genetic and Evolutionary Computation (MedGEC), pp. 2671–2678.

[49] Passaro, A., Baronti, F., Maggini, V., Micheli, A., Rossi, A.M. and Starita, A. 2005. Exploring relationships between genotype and oral cancer development through XCS. Proceedings of the 7th Annual Conference Companion on Genetic and Evolutionary Computation, Washington, D.C. Workshop session: Medical Applications of Genetic and Evolutionary Computation (MedGEC), pp. 147–151.

[50] Gotshall, S., Browder, K., Sampson, J., Soule, T. and Wells, R. 2007. Stochastic optimization of a biologically plausible spino-neuromuscular system model. *Genet. Progr. Evolv. Mach.* 8(4):355–380. doi:http://dx.doi.org/10.1007/s10710-007-9044-8.

[51] Howard, D.M., Tyrrell, A.M. and Cooper, C. 2007. Evolution of adult male oral tract shapes for close and open vowels. Proceedings of the 9th Annual Conference Companion on Genetic and Evolutionary Computation, London, UK. Workshop session: Medical applications of genetic and evolutionary computation (MedGEC), pp. 2751–2758.

[52] Stephens, C., Waelbroeck, H., Talley, S., Cruz, R. and Ash, A. 2005. Predicting healthcare costs using gas. Proceedings of the 7th Annual Conference Companion on Genetic and Evolutionary Computation, Washington, D.C. Workshop session: Medical Applications of Genetic and Evolutionary Computation (MedGEC), pp. 159–163.

[53] Petrovski, A. and McCall, J. 2005. Smart problem solving environment for medical decision support. GECCO '05. Proceedings of the 2005 Workshops on Genetic and Evolutionary Computation, Washington, D.C., 25–26 June. ACM, New York, pp. 152–158. doi:http://doi.acm.org/10.1145/1102256.1102290.

[54] Yeh, J. and Lin, W. 2007. Using simulation technique and genetic algorithm to improve the quality care of a hospital emergency department. *Expert Syst. Appl.* **32**(4):1073–1083. doi:http://dx.doi.org/10.1016/j.eswa.2006.02.017.

[55] Kim, T.-S., Lee, H., Park, J., Lee, C.-H., Lee, Y.-M., Choi, C.-S., Hwang, S.-G., Kim, H.-D. and Min, C.-H. 2008. Ubiquitous evolvable hardware system for heart disease diagnosis applications. *Int. J. Electron.* **95**(7): pp. 637–651.

[56] Goletsis, Y., Papaloukas, C., Fotiadis, D.I., Likas, A. and Michalis, L.K. 2004. Automated ischemic beat classification using genetic algorithms and multicriteria decision analysis. *IEEE Trans. Biomed. Eng.* **51**(10):1717–1725.

[57] Bhatia, S., Prakash, P. and Pillai, G.N. 2008. SVM based decision support system for heart disease classification with integer-coded genetic algorithm to select critical features. World Congress on Engineering and Computer Science (WCECS) 2008, International Association of Engineers, Hong Kong, pp. 34–38.

[58] Yan, H., Zheng, J., Jiang, Y., Peng, C. and Xiao, S. 2008. Selecting critical clinical features for heart diseases diagnosis with a real-coded genetic algorithm. *Appl. Soft Comput.* **8**(2):1105–1111. doi:http://dx.doi.org/10.1016/j.asoc.2007.05.017.

[59] Stylios, C.D. and Georgopoulos, V.C. 2008. Genetic algorithm enhanced fuzzy cognitive maps for medical diagnosis. FUZZ-IEEE 2008. IEEE International Conference on Fuzzy Systems (IEEE World Congress on Computational Intelligence), 1–6 June, pp. 2123–2128.

[60] Kim, H.T., Kim, B.Y., Park, E.H., Kim, J.W., Hwang, E.W., Han, S.K. and Cho, S. 2005. Computerized recognition of Alzheimer disease – EEG using genetic algorithms and neural network. *Future Gener. Comput. Syst.* **21**(7).

[61] Hazell, A. and Smith, S.L. 2008. Towards an objective assessment of Alzheimer's disease: the application of a novel evolutionary algorithm in the analysis of figure copying tasks. GECCO '08. Proceedings of the 2008 GECCO Conference Companion on Genetic and Evolutionary Computation, Atlanta, GA, 12–16 July. ACM, New York, pp. 2073–2080. doi:http://doi.acm.org/10.1145/1388969.1389024.

[62] Smith, S.L., Gaughan, P., Halliday, D.M., Ju, Q., Aly, N.M. and Playfer, J.R. 2007. Diagnosis of Parkinson's disease using evolutionary algorithms. *Genet. Progr. Evolv. Mach.* **8**(4):433–447.

[63] Gaál, B., Vassányi, I. and Kozmann, G. 2005. An evolutionary divide and conquer method for long-term dietary menu planning. *Lecture Notes in Computer Science*, Vol. 3581/2005, pp. 419–423.

[64] Blouza, A., Dumas, L. and M'Baye, I. 2008. Multiobjective optimization of a stent in a fluid-structure context. GECCO '08. Proceedings of the 2008 GECCO Conference Companion on Genetic and Evolutionary Computation, Atlanta, GA, 12–16 July. ACM, New York, pp. 2055–2060. doi:http://doi.acm.org/10.1145/1388969.1389021.

[65] Lameijer, E., Bäck, T., Kok, J.N. and Ijzerman, A.P. 2005. Evolutionary algorithms in drug design. *Nat. Comput. Int. J.* **4**(3):177–243. doi:http://dx.doi.org/10.1007/s11047-0045237-8.

4

Applications of GEC in Medical Imaging

4.1

Evolutionary Deformable Models for Medical Image Segmentation: A Genetic Algorithm Approach to Optimizing Learned, Intuitive, and Localized Medial-based Shape Deformation

Chris McIntosh and Ghassan Hamarneh
Medical Image Analysis Lab, Simon Fraser University, Burnaby, Canada

4.1.1 Introduction

Medical image segmentation remains a daunting task, but one whose solution will allow for the automatic extraction of important structures, organs and

Genetic and Evolutionary Computation: Medical Applications   Edited by Stephen L. Smith and Stefano Cagnoni
© 2011 John Wiley & Sons, Ltd

diagnostic features, with applications to computer-aided diagnosis, statistical shape analysis, and medical image visualization. Several classifications of segmentation techniques exist, including edge, pixel, and region-based techniques, in addition to clustering, and graph-theoretic approaches (Robb, 2000; Sonka and Fitzpatrick, 2000; Yoo, 2004). However, the unreliability of traditional, purely pixel-based methods in the face of shape variation and noise has caused recent trends (Pham *et al.*, 2000) to focus on incorporating prior knowledge about the location, intensity, and shape of the target anatomy (Hamarneh *et al.*, 2001). One method that has been of particular interest in meeting these requirements is deformable models, due to their inherent smoothness properties and ability to fit to ill-defined boundaries compared with, say, region-growing approaches.

Deformable models for medical image segmentation have gained popularity since the 1987 introduction of snakes by Terzopoulos and co-workers (Kass *et al.*, 1987; Terzopoulos, 1987). In addition to physics-based explicit deformable models (McInerney and Terzopoulos, 1996; Montagnat *et al.*, 2001), geometry-based implicit implementations also attracted attention (Caselles *et al.*, 1997; Osher and Paragios, 2003; Sethian, 1996). Several techniques were proposed to improve segmentation results by controlling model deformations (Cohen, 1991; Xu and Prince, 1998). However, with only smoothness and image-based constraints on their deformations, these models were highly susceptible to weak edges, noise, and entrapment in local minima (Figure 4.1.1).

In many applications, prior knowledge about object shape variability is available or can be obtained by studying a training set of shape examples. This knowledge restricts the space of allowable deformations to those that are anatomically feasible (Cootes *et al.*, 1995, 2001; Hamarneh and Gustavsson, 2000; Leventon *et al.*, 2000; Warfield *et al.*, 2000). One of the most notable works in this area is that of Cootes *et al.*, where they introduced and refined active shape models (ASM) (Cootes *et al.*, 1992, 1993, 1995). In ASM, principal component analysis (PCA) is calculated over a set of landmark points extracted from training shapes. The resulting principal components are used to construct a point distribution model (PDM) and an allowable shape domain (ASD). In a natural extension to their previous work, Cootes *et al.*, modify their method to include image intensity statistics (Cootes *et al.*, 2001). Staib and Duncan constrained the deformable models in Fourier space through conforming to probability distributions of the parameters of an elliptic Fourier decomposition of the boundary (Staib and Duncan, 1992). Statistical prior shape knowledge was also incorporated in implicit, level set-based deformable models. Leventon *et al.*, introduced statistical shape priors by using PCA to capture the main modes of variation of the level set function (Leventon *et al.*, 2000).

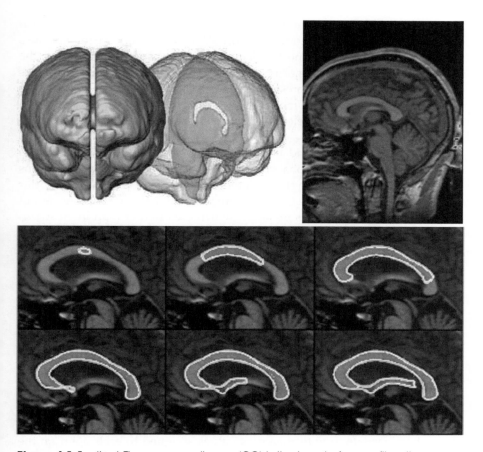

Figure 4.1.1 (top) The corpus callosum (CC) is the band of nerve fiber tissue connecting the left and right hemispheres of the brain. (bottom) Incorrect progress of a snake (white) segmenting the CC in an MRI image. Due to the proximity of nearby edges, manual initialization of the contour is required inside the CC. Even with proper initialization, leaking occurs due to the weak edges at the tip of the rostrum. Both of these problems are addressed by our proposed method's use of genetic algorithms (GAs) with prior shape knowledge

More recently, non-linear shape models have been introduced to address the problems that occur in practice due to linear approaches such as PCA (Cootes and Taylor, 1997; Dambreville *et al.*, 2006; Etyngier *et al.*, 2007; Sozou *et al.*, 1995). However, we argue the problem is not necessarily due to the application of a linear model to non-linear data but rather due to the global aspect of the shape statistics themselves, and due to the shape representation to which they are applied. Global shape statistics are those that model the shape variation of the entire shape simultaneously, i.e. each shape is a single point in some high-dimensional space, and the statistics (linear or not) describe some restricted set of that space. In terms of the shape representation, the majority of previous deformable model approaches adopt a boundary-based representation. As a consequence, the statistics are calculated using boundary models of the shape instead of models representing the interior and skeletal topology of the structures, and follow its natural geometry.

Global shape statistics and boundary-based shape representations both have inherent problems. Global statistics are unable to restrict deformations to particular locations of the shape model and unable to enforce shape prior only at selected local regions. Consequently, these global deformations cannot adapt to localized variations in shape, which are often of high interest, e.g. regions of pathology. The deformations resulting from boundary-based shape models are non-intuitive, i.e. they do not respect the underlying geometry of the object, and are not properly spatially constrained, i.e. their extent is not spatially localized (to a location or scale). The resulting statistically and boundary-based deformations themselves are un-intuitively defined as it is unclear what deforming along a particular mode of variation (even if localized) will accomplish; whether it will bulge, stretch, or bend the shape or a combination of those at multiple locations.

The solution is to use localized shape statistics over a medial-based shape representation, allowing deformations to be further quantified into specific types (bending, bulging, stretching) that are intuitive to the end user. Specifically, medial axis-based 2D shape representations enable such deformations by describing the object's shapes in terms of an axis positioned in the middle of the object along with thickness values assigned to each point on the axis that imply the shape of the boundary. Medial shape representations have been emerging as a powerful alternative (Blum, 1973; Dimitrov *et al.*, 2003; Hamarneh *et al.*, 2004; Hamarneh and McIntosh, 2005; Kaleem and Stephen, 2008; Pizer *et al.*, 2003; Sebastian *et al.*, 2001; Siddiqi *et al.*, 2002) to boundary-based techniques. Statistics on the non-linear manifolds of medial representations have also been proposed (Fletcher *et al.*, 2004).

Furthermore, there are numerous issues common to all of the aforementioned deformable model-based techniques (explicit, implicit, and statistically constrained). The first is that the model still requires initialization at some suitable target location of an image, with some shape, orientation, and scale. Another issue common to all of these deformable model-based methods is that the fitting to image data is typically performed through the minimization of a particular energy functional or some higher-level, for example user-driven, control process (Hamarneh *et al.*, 2001).

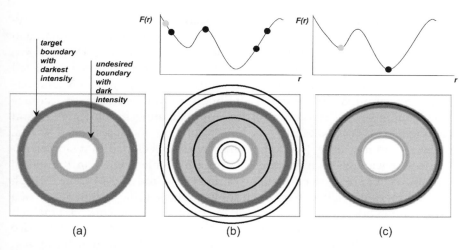

Figure 4.1.2 Synthetic toy example of single-parameter deformable model with local minima. The circular deformable model's only parameter is its radius r. The energy function $F(r)$ is minimal for the circle with darkest average intensity. The input image is shown in (a) with the darkest outmost boundary representing the global minima. In (b) traditional deformable models are initialized at a single point shown as light gray, while GA-based statistical deformable models are initialized at multiple locations (shown in black) and perform mutations on r. In (c) the GA converges to the global minima (darkest), while the deformable model gets stuck in a brighter local minima

Energy-functional minimization can be carried out in a variety of ways. One solution is to perform explicit differentiation under the Euler–Lagrange constraints, where each new application with a modified energy functional must be accompanied by one such derivation. The result is a set of constraints, which, if satisfied, guarantee an inflection point of the energy functional. The solution is then obtained through a gradient descent process where the change in the shape model (with respect to an artificial time variable) is equated to the Euler–Lagrange constraint, i.e. the deformable model comes to rest when the constraints are satisfied. However, the number of dependent variables (shape, location, scale, orientation, etc.) increases with the complexity of the search space, which often increases the number of local minima (Figure 4.1.2) and requires the calculation of an increasingly large number of derivatives. Therefore, iterative optimization of local minima-laden energy functionals is an important issue in medical image segmentation.

As such, there have been a number of recent approaches to obtaining the global optima of energy functionals (Boykov and Funka-Lea, 2006; Boykov and Kolmogorov, 2003; Grady, 2006; Nikolova, *et al.*, 2006). Graph cuts were demonstrated as a global minimization technique for a popular energy-functional (Caselles *et al.*, 1997), that computes a geodesic on a Riemannian space whose metric is computed from the image (Boykov and Kolmogorov, 2003). However, graph cuts have been shown to apply only to a restricted class of energy functionals (Kolmogorov and Zabin, 2004), and their solutions are discrete approximations to the continuous formulations whose accuracy is dependent on the resolution of the approximating graph (Boykov and Kolmogorov, 2003). Naturally, as that resolution increases, so does the method's running time. Random walkers were developed in a similar nature, solving image segmentation as a graph problem wherein the global optimum is obtained to a particular cost function (Grady, 2006). In fact, graph cuts and random walkers have been shown to be specific instantiations of a single framework (Sinop and Grady, 2007). Another line of work has emerged from the relaxation of the underlying shape model from a non-convex space to a convex one, thereby defining convex energy functionals which can then be minimized instead of their non-convex counterparts. The work began in 2006 with a simple restricted class of functionals (Nikolova *et al.*, 2006), which was then extended to a broader class (Bresson *et al.*, 2007), and then finally a similar work appeared in 2008 with the addition of a shape prior (Cremers *et al.*, 2008). However, restrictions still exist in that the functionals and the shape priors must be convex when defined over the relaxed space, and that the relaxed shape space must itself be convex. As a consequence, these works are not a general solution to the problem.

Therefore, what is required is a method that generalizes to a larger class of functionals and avoids calculating or estimating the shape model's evolution equation, while allowing the exploration of the search space in a manner that still converges towards an optimal solution. GAs are an example of one such method that retain speed by avoiding gradient calculations, allow the exploration to be carried out from a variety of initial locations, and reflect the learned variations of shape in terms of bends, bulges, and stretches.

Our proposed method utilizes prior knowledge to produce feasible deformations while also controlling the scale and location of these deformations. Moreover, through the use of GAs, we address the initialization problem, improve resistivity to local minima, and allow the optimization of highly customizable energy functionals, while avoiding costly derivative estimations. GAs allow us to explore a large number of simultaneous solutions in a very high-dimensional search space. We take advantage of all these properties of GAs and, at the same time, decompose the shape deformations into intuitive and localized constituents (bulge, bend, and stretch) that render the results more interpretable by clinicians (e.g. how much bend was needed in a particular part to fit to the new patient anatomy).

Other works have used GAs to drive traditional deformable models (Ballerini, 1998, 2001; MacEachern and Manku, 1998). Ballerini extends the classical active contour models (Terzopoulos, 1987) by using GAs to directly minimize the standard energy functional (Ballerini, 1998). Members of the GA population are hypothetical shape configurations, represented by their explicit contour locations. The method was later extended to color images by using one image energy-cost term per color channel (Ballerini, 2001). MacEachern and Manku presented a similar method using a binary representation of the contour (MacEachern, 1998). However, these methods are based on classical active contour models without incorporating prior shape knowledge (aside from simple smoothness constraints), which makes them prone to latching on to erroneous edges and ill-equipped to handle gaps in object boundaries. In Hill and Taylor (1992), GAs were used with statistically based ASMs, where the parameter space consists of possible ranges of values for the pose and shape parameters of the model. The objective function to be maximized reflects the similarity between the gray levels related to the object in the search stage and those found from training. Although this technique applies GAs to produce generations of plausible populations of shapes, the statistically based deformations are global (over the whole shape) and may not offer the required flexibility to accommodate for shape variations that are restricted to particular locations, nor are they intuitively defined (e.g. bulge, bend, stretch) deformations.

Our contribution in this chapter is a new segmentation method that uses GAs to address the typical initialization and local minima problems associated with traditional energy-minimization techniques, while maintaining a statistically feasible shape. Furthermore, the use of a medial-based shape representation provides an intuitive way to control deformations, while localized statistics restrict the variations and deformations to specific anatomical regions of a shape. To the best of our knowledge, none of the existing image segmentation techniques evolve a population of shapes using intuitive, spatially-constrained, and plausible deformations, nor have they employed fitness functions for GAs that are customizable for the problem domain.

In the following, we develop our method for medical image segmentation using GAs to drive statistically based, controlled (location and scale) and intuitive (bend, bulge, and stretch) deformations towards optimal solutions of a problem-specific energy functional. We describe how to obtain statistically constrained and intuitive deformations (Section 4.1.1.1), resulting in a shape representation that allows for statistically constrained deformation types at specific locations and scales. In Section 4.1.1.2, we present an overview of GAs. We detail how we employ GAs to drive our statistically based deformations in Section 4.1.2. Finally, we present results in Section 4.1.3, and conclude in Section 4.1.4.

4.1.1.1 Statistically Constrained Localized and Intuitive Deformations

We use our multi-scale (hierarchical) and multi-location (regional) PCA method introduced in Hamarneh *et al.* (2004) on a training set of medial shape profiles computed using 46 mid-sagittal CC images (Shenton *et al.*, 1992). We will first give an overview of medial shape profiles and then proceed to describe how hierarchical regional principal component analysis (HRPCA) can be applied.

Medial axis-based 2D shape representations enable deformations by describing the object's shape in terms of an axis positioned in the middle of the object along with thickness values assigned to each point on the axis that imply the shape of the boundary. We therefore describe the shapes as a mapping $x : \mathbb{R} \to \mathbb{R}^4$, the domain of which is a parameter m that traverses the medial axis. Although we use a single primary medial axis in this work, secondary medial axes are needed to represent more complex structures. The range of the mapping consists of four scalar values for each m, constituting four medial profiles. These are a length profile $L(m)$, an orientation profile $R(m)$, a left (with respect to the medial axis) thickness profile $T^l(m)$, and a right thickness profile $T^r(m)$, where $m = 1, 2, \ldots, N$; N is the number of medial nodes, and nodes 1 and N are the terminal nodes. The length profile represents the distances between consecutive pairs of medial nodes, and the orientation profile represents the angles between segments connecting consecutive pairs of medial nodes. The thickness profile represents the distances between medial nodes and their corresponding boundary points on both sides of the medial axis (Figure 4.1.3, bottom). Corresponding boundary points are calculated by computing the intersection of a line passing through each medial node in a direction normal to the medial axis, with the boundary representation of the object. Example medial profiles are shown in Figure 4.1.3 (top).

These profiles are rotation- and translation-invariant and capture intuitive measures of shape: length, orientation, and thickness. Reconstructing the shape after altering these profiles produces intuitive, controlled deformations: stretching, bending, and bulging, respectively.

In HRPCA, the principal component analysis is a function of the location, scale, and type of shape profile (length, orientation, or thickness) (Figure 4.1.4). Hence we obtain an average medial sub-profile, the main modes of

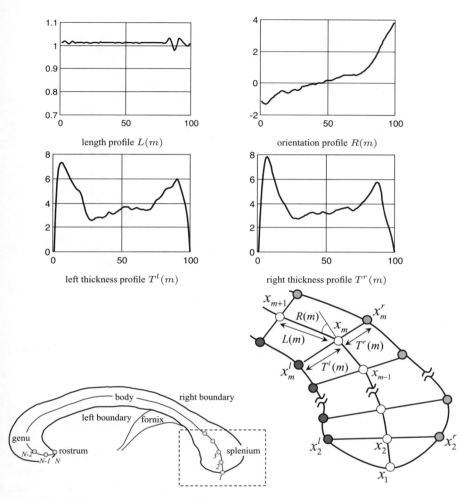

Figure 4.1.3 (top) Example medial shape profiles used to reconstruct the CC. (middle) Anatomically labeled CC shape reconstruction resulting from the medial profiles. (bottom) Details of outlined reconstruction showing medial profiles shape representation. Medial nodes shown as white disks, left and right boundary nodes shown in dark and light gray, respectively. x_m, x_m^l and x_m^r are the mth medial, left boundary and right boundary nodes, respectively. $L(m)$, $R(m)$, $T^l(m)$ and $T^r(m)$ are the length, orientation, left and right thickness profile values, respectively

Source: Adapted from Hamarneh *et al.* (2004)

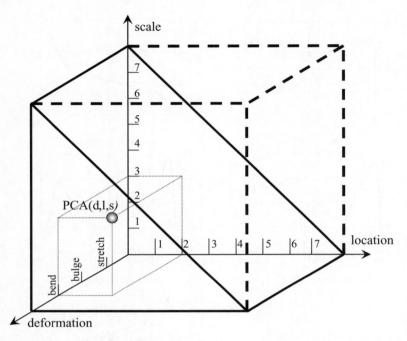

Figure 4.1.4 Hierarchical regional principal component analysis is a function of the deformation (d), location (l), and scale (s)

Source: Adapted from Hamarneh *et al.* (2004)

variation, and the amount of variation each mode explains for each location (*loc*), scale (*scl*), and shape profile type.

Global PCA becomes a special case of HRPCA by specifying $loc = 1$ and $scl = N$, where N is the number of shape variables covering the whole extent of the object. Hence we obtain N modes of variation of length N. In general, setting $loc = l$ and $scl = s$ will produce $s \times 1$ values, say thickness values for the T_r profile, and, as such, results in an $s \times s$ covariance matrix for the s modes of variation of dimensionality s. Consequently, we can now generate a statistically feasible stretch, bend, or bulge deformation at a specific location and scale in terms of the corresponding main modes of variation.

Generally with HRPCA, for a single deformation, location, and scale-specific PCA, we obtain the following model of medial profile variations:

$$p_{dls} = \bar{p}_{dls} + M_{dls} w_{dls} \qquad (4.1.1)$$

where p is the shape profile, d is the deformation profile type, l and s are the location and scale values of the deformation, $\bar{p}_{dls}$ is the average medial profile, M_{dls} describes the main variation modes, and w_{dls} are weights of the variation modes and are typically limited to ± 3 standard deviations. We note that in the case of our orientation profile the statistics are approximated under a vector space. However, there has been work on principle geodesic analysis that effectively performs the equivalent of PCA but intrinsic to the manifold of the medial shape representation (Fletcher 2004).

Note that for any shape profile type multiple variation modes can be activated by setting the corresponding weighting factors to non-zero values, according to

$$p_d = \bar{p}_d + \sum_l \sum_s M_{dls} w_{dls}. \tag{4.1.2}$$

In summary, varying the weights of one or more of the variation modes alters the length, orientation, or thickness profiles and generates, upon reconstruction, statistically feasible stretch, bend, or bulge deformations at specific locations and scales.

4.1.1.2 Genetic Algorithms

GAs are a special form of local search that model our understanding of evolution. In essence, a number of simultaneous agents (the population) each having an encoded state (the chromosome) perform a random walk (mutations) around the search space, while forming new solutions from combinations of existing solutions (crossover) and, thus, adjusting and refocusing the efforts of the search on exceptionally good areas once located. A few important choices are made during any application of genetic algorithms, involving how to encode the population (binary, integer, decimal, etc.), how to mutate the population (mutate all genes, some genes, etc.), how to select the parents for crossovers (roulette wheel, tournament selection), how to perform those crossovers (uniform, single-point), and, finally, what objective function to use for evaluating the fitness of the members of the population. Though these choices seem complex, in situations where the energy functional has hundreds or even thousands of dependent variables and parameters these few choices can yield nearly optimal values for all unknown variables and parameters concerned.

4.1.2 Methods

In this work, we use GAs to address the typical initialization, local minima and parameter sensitivity problems associated with traditional energy-minimization techniques. Moreover, the medial shape representation provides an intuitive way to synthesize and control deformations, while HRPCA enables localized statistics, thereby localizing the variations and deformations to specific anatomical regions of a shape (Figure 4.1.3, middle).

In this section we describe: our representation of individuals, our encoding of the model into chromosomes (deformation weights) to be optimized, our method of mutating (deforming) the model, and our fitness function (energy functional to minimize).

4.1.2.1 Population Representation

In medical image segmentation using GAs, the individuals forming the population represent potential shapes of the target structure, each having some level of accuracy measured by the fitness function (Section 4.1.2.4). Consequently, we require a shape representation consistent across models and capable of intuitive mutations. One straightforward way of representing shapes is using boundary nodes. However, intuitive aspects of shape variation (such as bending, thickness, and elongation) can not be easily captured and, therefore, not properly represented in the mutations, selection, and crossover phases. We require a shape representation that allows us to describe and control the shape deformations intuitively and in terms of our calculated shape statistics. Consequently, we represent each individual by its associated stretching, bending, and thickness profiles along with its global orientation, base location, and scale (Section 4.1.1.1).

4.1.2.2 Encoding the Weights for GAs

We use chromosomes to represent the set of weights of the principal components as obtained from the HRPCA, where each gene represents a weight (as a floating point number) for a particular deformation, location, scale, and mode of variation (Figure 4.1.5, top). In total there are at most $4 \times N \times (N - l + 1) \times s$ weights available for mutation since for each of the four deformations, d, we have N different locations, but for each location, l, we can only have up to $N - l + 1$ scales, s, each of which has at most s weights for the s principal components. In practice, only a subset of the s variation modes are required to explain a large percentage of the variance (Figure 4.1.4). In our

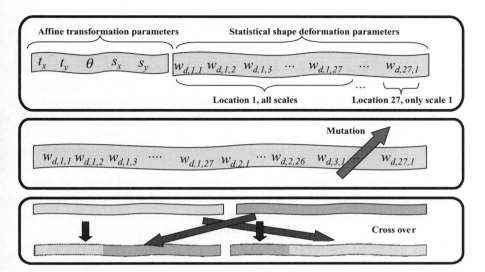

Figure 4.1.5 Segmenting an anatomical structure amounts to finding the optimal set of shape parameters. In our GA implementation (Section 4.1.2.1) we represent each shape as a chromosome with genes encoding affine and statistical shape deformation parameters (top). Mutation (Section 4.1.2.3) is performed by altering the weights of the HRPCA (middle). Crossover amounts to swapping a set of weights between two individuals (bottom)

application, $N = 27$, adding up to 14,616 dependent variables for our model, which motivates the use of GAs to search the very high-dimensional space.

4.1.2.3 Mutations and Crossovers

As previously discussed, GAs use mutations to walk randomly around the search space and crossovers to initialize search in new locations (locations that exist as some combination of pairs of current solutions).

With regard to mutations, in order to steer the evolution in a coarse to fine approach and, thereby, facilitate faster initial convergence, we initially constrain the mutations to the affine transformation parameters: base node position (translation values) (t_x, t_y), model orientation angle θ, and scale values (s_x, s_y) (Figure 4.1.5, top). Since our initial shape is the mean CC (obtained by setting all weights, w_{dls}, to zero), it can be expected to provide a reasonably strong fitness value when an acceptable position, orientation, and scale are set. In essence, we eliminate the possibility of getting a low score for a good location, scale and orientation, simply because of a bad random shape mutation.

With an adequate location, scale, and orientation obtained, we allow the mutations to produce shape deformations (Figure 4.1.5, middle). Mutation of a single gene amounts to altering its corresponding weight by sampling it from a uniform probability distribution under the constraint that the resulting weight lies within ±2 standard deviations of the corresponding mode of variation (i.e. ±2 times the square root of the explained variance obtained in HRPCA). Modifying a weight will change the medial profiles and hence the reconstructed shape boundary (Section 4.1.2.4).

During the initial phases of the evolution, every member of the population undergoes a random deformation with global scale, and at random amplitudes set in multiples of the corresponding standard deviations, thus resulting in a new shape. The initial constraint to global deformations is well-suited for our statistical deformations as localized deformations (say bulging the splenium) will not help until an acceptable global fit is obtained. Consequently, after a set number of generations has passed, we allow the deformations to begin varying in both position and scale to include, at first, larger deformations (those corresponding to an entire anatomical region, and, hence, a primary area of variation), and then, smaller deformations, which amounts to small variations in local regions.

In essence, GAs use crossover to combine the information from two existing "*parents*" into a single "*offspring*", that contains genes from each parent. We used uniform crossover, which makes an independent random decision for each gene whereby both parents have an equal probability of making the contribution.

4.1.2.4 Calculating the Fitness of Members of the GA Population

Our fitness function is specifically designed for segmentation of the CC, though, as noted in Section 4.1.1, the use of GAs allows us to easily adapt the function to any given task, including both prior shape and image-based knowledge. This is something traditional deformable models are not amenable to and globally optimal energy-minimizing models do not allow for (Kolmogorov and Zabin, 2004). For example, we have adopted a fitness function of the member shape i, $Fit(i)$, which considers the mean and standard deviation of the image intensity enclosed by the shape's boundary, Ω, the average edge strength along Ω, and the anatomical size of the CC. In particular we use

$$Fit(i) = \alpha S(i) + \beta \left(1 - e^{\left(\frac{-E(i)}{\chi}\right)}\right) + \phi \left(1 - e^{\left(\frac{-\eta(i)}{\varphi}\right)}\right) + \delta e^{\left(\frac{-\psi(i)}{\varepsilon}\right)} \qquad (4.1.3)$$

where

$$S(i) = \frac{|\Omega_{internal}|}{\bar{A}} \tag{4.1.4}$$

$$E(i) = \frac{1}{|\Omega|} \int_{\Omega} \|\nabla I\| \tag{4.1.5}$$

$$\eta(i) = \frac{1}{|\Omega_{internal}|} \int_{\Omega_{internal}} I \tag{4.1.6}$$

$$\psi(i) = \sqrt{\frac{1}{|\Omega_{internal}|} \int_{\Omega_{internal}} (I - \eta(i))^2}.$$

$\Omega_{internal}$ is the space enclosed by the reconstructed boundary, $|\Omega_{internal}|$ is the area of the space enclosed by the reconstructed boundary (Section 4.1.2.4), $|\Omega|$ is the length of the boundary, and $\bar{A}$ is the average size of the CC, learned from the training set. I is the image of the CC being segmented, and $\chi, \varepsilon, \varphi$ are learned edge strength, standard deviation, and mean intensity from the training data. Hence, $S(i)$ represents the area of shape i, $E(i)$ the average gradient magnitude along the shape's boundary, $\eta(i)$ the mean image intensity enclosed by Ω, and $\psi(i)$ the standard deviation of the image intensity enclosed by Ω. Finally, α, β, ϕ, and δ are scalar weights controlling the importance of each term in the segmentation process. The form $1 - e^x$ in the second and third terms favors high edge gradient along Ω, and high mean intensity of the segmented CC, whereas the form e^x in the fourth term favors homogeneous intensity (i.e. a small standard deviation).

Shape Reconstruction for Fitness Calculation

In order to evaluate the fitness, the boundary of the shape, Ω, must be reconstructed from the set of affine parameters and medial profiles specified by the shape weights. To reconstruct the object's shape given its set of medial profiles, we calculate the positions of the medial and boundary nodes from a known reference node at location $x_1 = (t_x, t_y)$. The next node at position $x_2 = x_1 + v_1$ is specified using an offset vector v whose angle is specified by the orientation profile plus the base angle θ, and whose length is specified by the stretch profile scaled by (s_x, s_y). The corresponding boundary nodes x_2^l and x_2^r (Figure 4.1.3, bottom) are then orthogonal to the medial axis, at a distance specified by the thickness profile scaled by (s_x, s_y). This process is repeated recursively, generating $x_3 = x_2 + v_2$, and so on. For details see Hamarneh *et al.* (2004). Finally, with the medial profiles like those shown in Figure 4.1.3 (top) as an input, we can reconstruct the CC structure in Figure 4.1.3 (bottom).

4.1.3 Results

We validate our method through the segmentation of the CC, which is the largest white-matter tract in the human brain. Specifically, it serves as the primary means of communication between the two cerebral hemispheres and mediates the integration of cortical processes from opposite sides of the brain. The presence of morphological differences in the corpus callosum in schizophrenics has been the subject of intense investigation (Rosenthal and Bigelow, 1972). The CC may also be involved in Alzheimer's disease (Pantel *et al.*, 1999), mental retardation (Marszal *et al.*, 2000), and other neurological disorders.

We present qualitative as well as quantitative results of the fully automatic segmentation of 46 CC in mid-sagittal MRI (Shenton *et al.*, 1992) using our GA-driven, statistically constrained deformable models (Figures 4.1.6, 4.1.7). We compare our results to those previously obtained in Hamarneh and McIntosh (2005), where statistically constrained, physically based deformations are controlled by a CC-specific hand-crafted schedule and initialization method (Table 4.1.1). Here our use of GAs enabled us to obtain superior accuracy without depending on an application-specific schedule, but rather a simple fitness function; making our method more accurate and more extensible, e.g. as in Hamarneh *et al.* (2007).

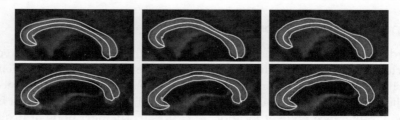

Figure 4.1.6 Two example segmentation results progressing left to right, showing fittest individual after automatic initialization (left), global deformations (middle), and local deformations (right). The white curve is the medial axis

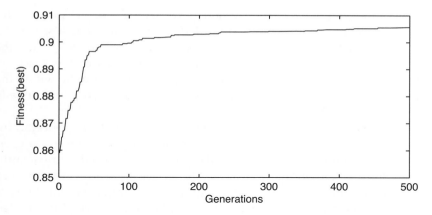

Figure 4.1.7 Plot of best individual fitness versus generation number

4.1.4 Conclusions

We have developed a novel segmentation technique by addressing the main concerns associated with both traditional and statistically based deformable models. Firstly, by using GAs to address the initialization, local minima, and parameter-sensitivity problems associated with traditional energy-minimization techniques. Secondly, our medial shape representation provides a powerful way to synthesize and analyze deformations, thus, decomposing deformations into different types that are intuitively controlled and are more easily communicated to medical experts than boundary-based deformations. Finally, our use of HRPCA enables localized statistics, thereby localizing the variations and deformations to specific anatomical regions, which render the results more interpretable by clinicians and enable regional statistical analysis.

Table 4.1.1 Segmentation error comparison

Error	mean	median	min	max	std
Hand-crafted (Hamarneh and McIntosh, 2005)	0.1834	0.1706	0.1095	0.4526	0.0576
GA	0.1719	0.1501	0.0732	0.5464	0.0868

Error $\varepsilon = (S \cup M - S \cap M)/M$ is used, where S and M denote the area enclosed within the result of the automatic segmentation and the manual expert delineation, respectively.

Furthermore, we have demonstrated how GAs can be combined with constrained shape deformations to effectively explore the search space of a complex energy functional, thereby incorporating prior knowledge into the solution while retaining multiple simultaneous searches of the space. In essence, we have constrained the random walks of the GA to lie within the allowable shape domain, thus greatly reducing the search space traditionally associated with deformable models.

Our method is also extensible to other segmentation problems. Specifically, given a training set of shapes for a different anatomical structure, one can perform skeletonization followed by medial profile extraction and, subsequently, HRPCA. Further, the components of the fitness functions presented here can apply to other anatomical structures, with possible minor modifications as the application warrants them. Nevertheless, other terms can easily be added related to texture, color, or other image features. Finally, we are working on extending these ideas to 3D, where the genes become the weights of 3D shape representation parameters.

Though other works have used GAs to drive deformable models (Ballerini, 1998, 2001; Hill and Taylor, 1992; MacEachern and Manku, 1998), to the best of our knowledge, no other works have combined GAs with statistically based deformations in a way that yields intuitively constrained deformations, nor have they employed fitness functions well suited to the problem domain. Furthermore, by comparison with iterative, gradient-descent approaches, our method retains speed by avoiding gradient calculations, allows search space exploration to be carried out from a variety of initial locations, and enables it to be done in a way that intuitively reflects the learned variations of shape (bends, bulges, and stretches).

Acknowledgments

We wish to thank Dr Martha Shenton of the Harvard Medical School for providing the MRI data, and Peter Plett for assisting with code development.

References

Ballerini L 1998 Genetic snakes for medical image segmentation. *Proc. SPIE, Mathematical Modeling and Estimation Techniques in Computer Vision* **3457**, 284–295.

Ballerini L 2001 Genetic snakes for color images segmentation. *Lecture Notes in Computer Science* **2037**, 268–277.

Blum H 1973 Biological shape and visual science. *Theoretical Biology* **38**, 205–287.

Boykov Y and Funka-Lea G 2006 Graph cuts and efficient n-d image segmentation. *International Journal of Computer Vision (IJCV)* **70**(2), 109–131.

Boykov Y and Kolmogorov V 2003 Computing geodesics and minimal surfaces via graph cuts. *ICCV '03: Proceedings of the Ninth IEEE International Conference on Computer Vision*, Vol. 1, pp. 26–33.

Bresson X, Esedoglu S, Vandergheynst P, Thiran JP and Osher S 2007 Fast global minimization of the active contour/snake model. *Journal of Mathematical Imaging and Vision* **28**(2), 151–167.

Caselles V, Kimmel R and Sapiro G 1997 Geodesic active contours. *International Journal of Computer Vision* **22**(1), 61–79.

Cohen LD 1991 On active contour models and balloons. *Computer Vision Graphics and Image Processing: Image Understanding* **53**(2), 211–218.

Cootes T and Taylor C 1997 A mixture model for representing shape variation. *British Machine Vision Conference*, pp. 110–119.

Cootes T, Taylor C, Cooper D and Graham J 1992 Training models of shape from sets of examples. *British Machine Vision Conference*, pp. 9–18.

Cootes T, Taylor C, Hill A and Halsam J 1993 The use of active shape models for locating structures in medical images. *Proceedings of the 13th International Conference on Information Processing in Medical Imaging*, pp. 33–47.

Cootes TF, Cooper D, Taylor CJ and Graham J 1995 Active shape models – their training and application. *Computer Vision and Image Understanding* **61**, 38–59.

Cootes TF, Edwards GJ and Taylor CJ 2001 Active appearance models. *IEEE Transactions on Pattern Analysis and Machine Intelligence* **23**(1), 681–685.

Cremers D, Schmidt FR and Barthel F 2008 Shape priors in variational image segmentation: Convexity, Lipschitz continuity and globally optimal solutions. *IEEE Conference on Computer Vision and Pattern Recognition (CVPR)*, Anchorage, Alaska.

Dambreville S, Rathi Y and Tannenbaum A 2006 Shape-based approach to robust image segmentation using kernel PCA. *IEEE Conference on Computer Vision and Pattern Recognition (CVPR)*, Vol. 1, pp. 977–984.

Dimitrov P, Damon J and Siddiqi K 2003 Flux invariants for shape. *IEEE Conference on Computer Vision and Pattern Recognition (CVPR)*, pp. 835–841.

Etyngier P, Segonne F and Keriven R 2007 Shape priors using manifold learning techniques. *International Conference on Computer Vision (ICCV)*, pp. 1–8.

Fletcher P, Lu C, Pizer S and Joshi S 2004 Principal geodesic analysis for the study of nonlinear statistics of shape. *IEEE Transactions on Medical Imaging* **23**(8), 995–1005.

Grady L 2006 Random walks for image segmentation. *IEEE Transactions on Pattern Analysis and Machine Intelligence* **28**(11), 1768–1783.

Hamarneh G, Abu-Gharbieh R and McInerney T 2004 Medial profiles for modeling and statistical analysis of shape. *International Journal of Shape Modeling* **10**(2), 187–209.

Hamarneh G and Gustavsson T 2000 Statistically constrained snake deformations. *IEEE International Conference on Systems, Man, and Cybernetics*, Vol. 3, pp. 1610–1615.

Hamarneh G and McIntosh C 2005 Physics-based deformable organisms for medical image analysis. *Proceedings of SPIE Medical Imaging: Image Processing*, Vol. 5747, pp. 326–335.

Hamarneh G, McInerney T and Terzopoulos D 2001 Deformable organisms for automatic medical image analysis. *Medical Image Computing and Computer-Assisted Intervention (MICCAI)*, pp. 66–76.

Hamarneh G, Ward A and Frank R 2007 Quantification and visualization of localized and intuitive shape variability using a novel medial-based shape representation. *IEEE International Symposium on Biomedical Imaging (IEEE ISBI)*, pp. 1232–1235.

Hill A and Taylor C 1992 Model-based image interpretation using genetic algorithms. *Image and Vision Computing* **10**(5), 295–300.

Kaleem S and Stephen P 2008 *Medial Representations: Mathematics, Algorithms and Applications.* Springer-Verlag.

Kass M, Witkin A and Terzopoulos D 1987 Snakes: Active contour models. *International Journal of Computer Vision* **1**(4), 321–331.

Kolmogorov V and Zabin R 2004 What energy functions can be minimized via graph cuts? *IEEE Transactions on Pattern Analysis and Machine Intelligence* **26**(2), 147–159.

Leventon M, Grimson E and Faugeras O 2000 Statistical shape influence in geodesic active contours. *IEEE Conference on Computer Vision and Pattern Recognition (CVPR)*.

MacEachern L and Manku T 1998 Genetic algorithms for active contour optimization. *IEEE Proceedings of the International Symposium on Circuits and Systems*, Vol. 4, pp. 229–232.

Marszal E, Jamrow E, Pilch J *et al.* 2000 Agenesis of corpus callosum: Clinical description and etiology. *Journal of Child Neurology* **15**, 401–405.

McInerney T and Terzopoulos D 1996 Deformable models in medical image analysis: A survey. *Medical Image Analysis* **1**(2), 91–108.

Montagnat J, Delingette H and Ayache N 2001 A review of deformable surfaces: topology, geometry and deformation. *Image and Vision Computing* **19**(14), 1023–1040.

Nikolova M, Esedoglu S and Chan TF 2006 Algorithms for finding global minimizers of image segmentation and denoising models. *SIAM Journal on Applied Mathematics* **66**(5), 1632–1648.

Osher S and Paragios N 2003 *Geometric Level Set Methods in Imaging Vision and Graphics.* Springer-Verlag.

Pantel J, Schroder J, Jauss M *et al.* 1999 Topography of callosal atrophy reflects distribution of regional cerebral volume reduction in Alzheimer's disease. *Psychiatry Research* **90**, 180–192.

Pham DL, Xu C and Prince JL 2000 A survey of current methods in medical image segmentation. *Annual Review of Biomedical Engineering* **2**, 315–338.

Pizer S, Gerig G, Joshi S and Aylward SR 2003 Multiscale medial shape-based analysis of image objects. *Proceedings of the IEEE* **91**(10), 1670–1679.

Robb RA 2000 *Biomedical Imaging, Visualization, and Analysis.* Wiley-Liss Inc.

Rosenthal R and Bigelow L 1972 Quantitative brain measurements in chronic schizophrenia. *British Journal of Psychiatry* **121**, 259–264.

Sebastian T, Klein P and Kimia B 2001 Recognition of shapes by editing shock graphs. *International Conference on Computer Vision (ICCV)*, pp. 755–762.

Sethian JA 1996 Level set methods; evolving interfaces in geometry, fluid mechanics. *Computer Vision and Material Sciences*. Cambridge University Press.

Shenton M, Kikinis R, Jolesz F, Pollak S, LeMay M, Wible C *et al.* 1992 Abnormalities in the left temporal lobe and thought disorder in schizophrenia: A quantitative magnetic resonance imaging study. *New England Journal of Medicine* **327**, 604–612.

Siddiqi K, Bouix S, Tannenbaum A and Zucker S 2002 Hamilton–Jacobi skeletons. *International Journal of Computer Vision* **48**(3), 215–231.

Sinop AK and Grady L 2007 A seeded image segmentation framework unifying graph cuts and random walker which yields a new algorithm. *International Conference on Computer Vision (ICCV)*, IEEE Computer Society.

Sonka M and Fitzpatrick J 2000 *Handbook of Medical Imaging, Volume 2: Medical Image Processing and Analysis*. SPIE-International Society for Optical Engine.

Sozou P, Cootes T, Taylor C and Di Mauro E 1995 Non-linear point distribution modelling using a multi-layer perceptron. *British Machine Vision Conference*, pp. 107–116.

Staib LH and Duncan JS 1992 Boundary finding with parametrically deformable models. *IEEE Transactions on Pattern Analysis and Machine Intelligence* **14**(11), 1061–1075.

Terzopoulos D 1987 On matching deformable models to images. *Topical Meeting on Machine Vision, Technical Digest Series* **12**, 160–167.

Warfield SK, Kaus M, Jolesz FA and Kikinis R 2000 Adaptive, template moderated, spatially varying statistical classification. *Medical Image Analysis* **4**(1), 43–55.

Xu C and Prince J 1998 Snakes, shapes, and gradient vector flow. *IEEE Transactions on Image Processing* **7**(3), 359–369.

Yoo TS 2004 *Insight into Images: Principles and Practice for Segmentation, Registration, and Image Analysis*. Ak Peters Ltd, suite 230, 888 Worcester St. Wellesey, MA, USA.

4.2

Feature Selection for the Classification of Microcalcifications in Digital Mammograms using Genetic Algorithms, Sequential Search and Class Separability

Santiago E. Conant-Pablos, Rolando R. Hernández-Cisneros, and Hugo Terashima-Marín
Centro de Computación Inteligente y Robótica, Tecnológico de Monterrey, Monterrey, Mexico

4.2.1 Introduction

Breast cancer is one of the main causes of death in women, and early diagnosis is an important means to reduce the mortality rate. Mammography is one

---

Genetic and Evolutionary Computation: Medical Applications   Edited by Stephen L. Smith and Stefano Cagnoni
© 2011 John Wiley & Sons, Ltd

of the most common techniques for breast cancer diagnosis, and calcifications one of the findings that can be seen on mammograms. These are very small calcium deposits that can appear within the soft tissue of the breast. Calcifications, that appear as white dots on the mammogram, are not always a sign of breast cancer, but sometimes an indication of a precancerous condition. Calcifications are divided into two types. Macrocalcifications are larger deposits of calcium, not usually linked to breast cancer. Microcalcifications are typically 100 microns to several millimeters in diameter, may occur in clusters of three or more and usually appear in areas less than 1 cm^2, or in patterns such as circles and lines. These microcalcifications are associated with extra cell activity in breast tissue, and tight clusters of microcalcifications can indicate early breast cancer.

However, the predictive value of mammograms is relatively low, compared with biopsy. This low sensitivity [8] is caused by the low contrast between the cancerous tissue and the normal parenchymal tissue, the small size of microcalcifications and possible deficiencies in the image digitalization process. The sensitivity may be improved through having each mammogram checked by two or more radiologists, with the consequence of making the process more resource intensive. A viable alternative is replacing one of the radiologists by a computer system, giving a second opinion [1, 19].

The computer system should process the mammograms as digitized images for the detection of suspicious points as being microcalcifications. Then, microcalcifications have to be separated through a classifications process and grouped into clusters. Finally, the obtained microcalcification clusters have to be classified as indicative of benign (harmless) breast tissue or malignant, a sign of breast cancer.

We decided to use artificial neural networks (ANNs) to construct the classifiers for the recognition of individual microcalcifications and benign/malignant microcalcification clusters. ANNs have been used successfully for classification purposes in medical applications. Unfortunately, for an ANN to be successful in a particular domain, its architecture, training algorithm and the domain variables selected as inputs must be adequately chosen. Designing an ANN architecture is a trial-and-error process requiring the selection and tuning of several parameters that depend strongly on the features selected to describe the training data. The classification problem could involve too many features (variables), most of them not relevant at all for the classification process itself. Genetic algorithms (GAs) may be used to address both problems mentioned above, helping to obtain more accurate ANNs with better generalization abilities. An exhaustive review of evolutionary artificial neural networks (EANNs) is presented by Yao [20] and Balakrishnan and Honavar [2].

In particular, this chapter describes and compares three methods for selecting the most relevant features extracted from both individual microcalcifications and microcalcification clusters, which provide the inputs of two simple feedforward ANNs for their classification, with an expectation of improved accuracy. The first method works by ordering features according to a class separability metric and then selecting the most relevant ones as inputs to an ANN. The second method implements a forward sequential search algorithm that sequentially adds features to the ANN while its classification error decreases, and stops when this error starts to increase. Finally, the third method uses a GA to evolve good feature subsets for the classification tasks.

The remainder of the chapter is organized as follows. In the second section, the methodology is discussed. The third section deals with the experiments and the main results of this work. Finally, in the fourth section, the conclusions are presented, and some comments regarding future work are also made.

4.2.2 Methodology

The mammograms used in this project were provided by the Mammographic Image Analysis Society (MIAS) [18]. The MIAS database contains 322 images, all medio-lateral (MLO) view, digitized at resolutions of 50 microns/pixel and 200 microns/pixel. In this work, images with a resolution of 200 microns/pixel were used. The data has been reviewed by a consultant radiologist, and all the abnormalities have been identified and marked. The truth data consists of the location of the abnormality and the radius of a circle which encloses it. From the totality of the database, only 25 images contain microcalcifications. Among these 25 images, 13 cases are diagnosed as malignant and 12 as benign. Some related works have used this same database [6, 9, 11, 14]. The general procedure, as shown in Figure 4.2.1, receives digital mammograms as input, and consists of five stages: pre-processing, detection of potential microcalcifications (signals), classification of signals into microcalcifications, detection of microcalcification clusters, and classification of microcalcification clusters into benign and malignant.

4.2.2.1 Pre-processing

This stage has the aim of eliminating those elements in the images that could interfere in the process of identifying microcalcifications. A secondary goal is to reduce the work area only to the relevant region that exactly contains the breast.

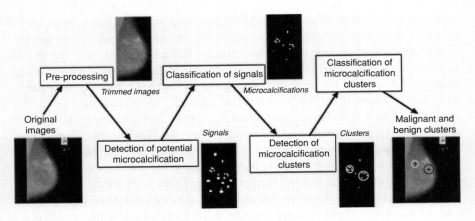

Figure 4.2.1 General procedure followed by the classification system

The procedure receives the original images as input. A median filter is applied in order to eliminate the background noise and then an automatic cropping procedure deletes the background marks and the isolated regions, so the image will contain only the region of interest. The result of this stage is a smaller image, with less noise.

4.2.2.2 Detection of Potential Microcalcifications (Signals)

The main objective of this stage is to detect the mass centers of the suspicious points as being microcalcifications in the images (signals). The pre-processed images of the previous stage are the inputs of this procedure. A method employing difference of gaussian filters (DoG) is used for enhancing those regions containing bright points. The use of DoG for detection of potential microcalcifications has been addressed successfully by Dengler, Behrens and Desaga [7] and Ochoa [15]. A gaussian filter is obtained from a gaussian distribution and when applied to an image, eliminates high-frequency noise, acting like a smoothing filter. The DoG filter is obtained from the difference of two gaussian functions, as shown in equation (4.2.1), where x and y are the coordinates of a pixel in the image, k_1 and k_2 are the heights of the functions, and σ_1 and σ_2 are the different standard deviations of the two gaussian filters that construct the DoG filter.

$$DoG(x, y) = k_1 \exp^{\left(\frac{x^2+y^2}{2\sigma_1^2}\right)} - k_2 \exp^{\left(\frac{x^2+y^2}{2\sigma_2^2}\right)} \qquad (4.2.1)$$

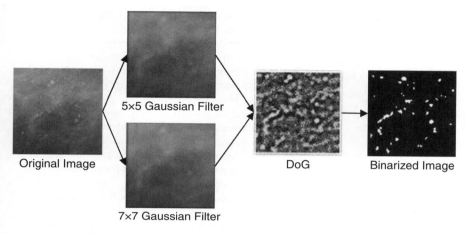

5×5 Gaussian Filter

Original Image   DoG   Binarized Image

7×7 Gaussian Filter

Figure 4.2.2 Example of the application of a DoG filter (5 × 5, 7 × 7)

The resultant image after applying a DoG filter is globally binarized, using a certain threshold. In Figure 4.2.2, an example of the application of a DoG filter is shown. A region-labeling algorithm allows the identification of each one of the points. Then, a segmentation algorithm extracts small 9 × 9 windows, containing the regions of interest whose centroids correspond to the centroids of the identified points. The size of the windows is adequate for containing the signals given that, at the current resolution of 200 microns, the potentially malignant microcalcifications have an area of 5 × 5 pixels on average [16].

In order to detect the greatest possible number of points, several DoG filters are applied several times, varying the binarization threshold. The points obtained by applying each filter are added to the points obtained by the previous one, deleting the repeated points.

All of the points obtained by the DoG filters are subsequently passed to three selection procedures in order to transform them into signals (potential microcalcifications). The first method performs selection according to the object area. The second method performs selection according to the gray level of the points. Finally, the third selection method uses the gray gradient (or absolute contrast, the difference between the mean gray level of the point and the mean gray level of the background). The result of these three selection processes is a list of signals (potential microcalcifications) represented by their centroids.

Table 4.2.1 Summary of features extracted from the signals (potential
microcalcifications)

Signal contrast (7 features)	Maximum gray level, minimum gray level, median gray level, mean gray level, standard deviation of the gray level, gray level skewness, gray level kurtosis.
Background contrast (7 features)	Background maximum gray level, background minimum gray level, background median gray level, background mean gray level, standard deviation of the background gray level, background gray level skewness, background gray level kurtosis.
Relative contrast (3 features)	Absolute contrast, relative contrast, proportional contrast.
Shape features (20 features)	Area, convex area, background area, perimeter, maximum diameter, minimum diameter, equivalent circular diameter, fiber length, fiber width, curl, circularity, roundness, elongation1, elongation2, eccentricity, aspect ratio, compactness1, compactness2, compactness3, solidity.
Contour sequence moments (7 features)	CSM1, CSM2, CSM3, CSM4, mean radii, standard deviation of radii.
Invariant moments of Hu (4 features)	IM1, IM2, IM3, IM4.

4.2.2.3 Classification of Signals into Microcalcifications

The objective of this stage of processing is to identify if an obtained signal corresponds to an individual microcalcification or not. With this in mind, the 47 features shown in Table 4.2.1 are extracted from each individual signal.

In order to process the signals and accurately classify the microcalcifications, we decided to use ANNs as classifiers. In the first section, we mentioned that one of the difficulties of working with conventional feedforward ANNs is that a classification problem could involve too many variables (features), and most of them may not be relevant at all for the classification process itself. For this reason, instead of using the full set of 47 features to construct the signal's classifier, we experimented and compared three methods for the selection of the subset of features to use: class separability, forward sequential search, and genetic algorithms.

Class Separability: As in [4] and [5] we decided to use the Kullback–Leibler (KL) distance between histograms of feature values to estimate how well a feature separates the data into different classes and to define an ordering of the 47 features. Two histograms are created for each feature, one for microcalcifications and another for other signals. For this, features are discretized using $b = \sqrt{|D|}$ equally spaced bins, where $|D|$ is the size of the training data. The histograms are later normalized by dividing the number of elements in each bin by the total number of elements, so the probability $p_j(d = i\,|\,n)$ that the jth feature takes a value in the ith bin of the histogram, given a class n, is obtained. For each feature j, the class separability Δ_j is calculated as

$$\Delta_j = \sum_{m=1}^{c}\sum_{n=1}^{c}\sum_{i=1}^{b} p_j(d = i\,|\,c = m)\log\left(\frac{p_j(d = i\,|\,c = m)}{p_j(d = i\,|\,c = n)}\right) \tag{4.2.2}$$

where c is the total number of classes, in our case two, one for microcalcifications and one for other signals. The features are finally ranked by sorting them in descending order of the distances Δ_j (larger distances mean better separability). In the application of this method, two features were heuristically assumed redundant if their distances differ by less than 0.0001, and the feature with the smallest distance is eliminated. Other irrelevant non-discriminative features with Δ_j distances less than 0.001 are eliminated also.

Forward Sequential Search: As in [17], we decided to use a method that implements a forward sequential search (FSS) algorithm [1].

A FSS algorithm starts with an empty feature set and in each iteration adds a new feature to the ANN used as the microcalcification's classifier while the error is decreasing, and stops when the error starts increasing again. The feature added in each iteration is chosen based on its ability to reduce the error in the classifier.

Our FSS algorithm uses the information gain of each feature as the ordering metric. The features are selected in the descending order of their gain values. To compute its information gain, the feature data is discretized, and the gain values are computed from the difference of entropy measures as

$$\text{gain}(S, A) = \text{entropy}(S) - \sum_{v \in \text{values of } (A)} \frac{|S_v|}{|S|}\text{entropy}(S_v) \tag{4.2.3}$$

where S are the signals to be classified, A is the feature for which the gain is computed, S_v is a subset of S where the A feature takes the v value, and the entropy function of a set of signals S is computed as

$$\text{entropy}(S) = \sum_{i \in C} -p_i(S, S_i) \log_2 p_i(S, S_i) \qquad (4.2.4)$$

where C is the set of class values, in our case if it is or is not a calcification, and p_i is the proportion of ocurrences of class i in the set S obtained as $p_i(S, S_i) = |S_i|/|S|$.

Genetic Algorithms: Expecting to achieve greater accuracy in the classification, we use a third method, this being based on a GA for selecting features. The chromosomes of the individuals in the GA contain 47 bits, one bit for each extracted feature; the value of the bit determines whether that feature will be used in the classification or not [3].

Each individual is evaluated by constructing and training a feedforward ANN (with a predetermined structure). The number of inputs of this ANN is determined by the subset of features to be included, coded in the chromosome. For solving nonlinearly separable problems it is recommended that at least one hidden layer is provided in the network, and according to Kolmogorov's theorem [13], considering the number of inputs (n), that the hidden layer contains $2n + 1$ neurons. The output layer has only one neuron. The accuracy of each network is used to determine the fitness of each individual.

When the GA stops, either because the generations limit has been reached or because improvements in the evaluation of the best individual have not been observed during five consecutive generations, the subset of features of the ANN with the best performance in terms of overall accuracy is obtained, and its ANN used as the microcalcification's classifier.

4.2.2.4 Detection of Microcalcification Clusters

During this stage, the microcalcification clusters are identified. The detection and posterior consideration of every microcalcification cluster in the images may produce better results in a subsequent classification process, as shown in [17]. Because of this, an algorithm for locating microcalcification cluster regions where the quantity of microcalcifications per square centimeter (density) is higher, was developed.

The basic clustering process starts by obtaining the set of all pairs formed with the detected microcalcifications, eliminating all pairs with a Euclidian distance between the centroids of its microcalcifications larger than an empirically defined threshold of 100 pixels. Then it enters a loop for creating the set of microcalcification clusters. In each iteration, the best new cluster, formed with microcalcifications that remain ungrouped, is added. For this, a list of neighbors for each remaining microcalcification is obtained containing all the remaining microcalcifications to a distance smaller than a chosen threshold. The density of each of these potential clusters is computed from its points count and the size of the area computed from the convex polygon it forms. Finally, the group with the maximum density is chosen as a new cluster, the microcalcifications that it contains are eliminated, and a new iteration is started. This process goes on until all the detected microcalcifications are added to a cluster. Every detected cluster is then labeled.

4.2.2.5 Classification of Microcalcification Clusters into Benign and Malignant

This stage has the objective of classifying each cluster into one of two classes: benign or malignant (information provided by the MIAS database). With this in mind, the 30 features shown in Table 4.2.2 are extracted from every microcalcification cluster detected in the previous stage.

In order to process microcalcification clusters and accurately classify them into benign or malignant, we decided again to use ANNs as classifiers. To determine which of the 30 extracted features from the clusters are relevant for their classification, we applied the same three methods described in a previous section: using the same class separability criteria, deriving a forward sequential search, and using genetic algorithms. The main difference in the implementation of the methods for classifying clusters resulted from their adaptation to the use of the 30 different features computed for the clusters.

4.2.3 Experiments and Results

4.2.3.1 From Pre-processing to Signal Extraction

As mentioned in the previous section, only 25 images from the MIAS database contain microcalcifications. Among these 25 images, 13 cases are diagnosed as malignant and 12 as benign. Three images were discarded because the positions of the microcalcification clusters, marked in the additional data that comes with the database, were outside the boundaries of the breast. So,

Table 4.2.2   Summary of features extracted from the microcalcification clusters

Cluster shape (14 features)	Number of calcifications, convex perimeter, convex area, compactness, microcalcification density, total radius, maximum radius, minimum radius, mean radius, standard deviation of radii, maximum diameter, minimum diameter, mean of the distances between microcalcifications, standard deviation of the distances between microcalcifications.
Microcalcification area (6 features)	Total area of microcalcifications, mean area of microcalcifications, standard deviation of the area of microcalcifications, maximum area of the microcalcifications, minimum area of the microcalcifications, relative area.
Microcalcification contrast (10 features)	Total gray mean level of microcalcifications, mean of the mean gray levels of microcalcifications, standard deviation of the mean gray levels of microcalcifications, maximum mean gray level of microcalcifications, minimum mean gray level of microcalcifications, total absolute contrast, mean absolute contrast, standard deviation of the absolute contrast, maximum absolute contrast, minimum absolute contrast.

only 22 images were finally used for this study, and they were passed through the pre-processing stage first (application of a median filter and trimming).

In the second phase, six gaussian filters of sizes $5 \times 5, 7 \times 7, 9 \times 9, 11 \times 11, 13 \times 13$, and 15×15 were combined, two at a time, to construct 15 DoG filters that were applied sequentially. Each one of the 15 DoG filters was applied 51 times to the pre-processed images, varying the binarization threshold in the interval [0, 5] in increments of 0.1. The points obtained by applying each filter were added to the points obtained by the previous one, deleting the repeated points. The same procedure was repeated with the points obtained by the remaining DoG filters. These points passed through the three selection methods for selecting signals (potential microcalcification), according to region area, gray level, and the gray gradient. For this work, a minimum area of 1 pixel (0.0314 mm^2) and a maximum of 77 pixels (3.08 mm^2) were considered. Studying the mean gray levels of pixels surrounding identified microcalcifications, it was found that they have values in the interval [102, 237] with a mean of 164. For this study, we set the minimum gray level for points to be selected to 100. Again, studying the mean gray gradient of points surrounding identified microcalcifications, it was found that they have values in the interval [3, 56] with a mean of 9.66. For this study, we set the minimum gray gradient for points to be selected to 3. The result was a list of 1,242,179 signals (potential microcalcifications) represented by their centroids.

The additional data included in the MIAS database define, with centroids and radii, the areas in the mammograms where microcalcification clusters are located. It is supposed that signals within these areas are mainly microcalcifications, but there are many signals that lie outside the marked areas. With these data and the support of expert radiologists, each of the signals located in these 22 mammograms was pre-classified as a microcalcification, or as non-microcalcification. Out of the 1,242,179 signals, only 4,612 (0.37%) were microcalcifications, and the remaining 1,237,567 (99.63%) were not. Because of this imbalanced distribution of examples of each class, an exploratory sampling was made. Several samplings with different proportions of each class were tested and finally we decided to use a sample of 10,000 signals, including 2,500 microcalcifications (25%).

4.2.3.2 Classification of Signals into Microcalcifications

After the 47 microcalcification features were computed from each signal, the first method for feature selection, based on class separability for ranking the features, reduced the relevant features to five: median gray level, mean gray level, minimum gray level, background maximum gray level, and background mean gray level. A transactional database was obtained, containing 10,000 signals (2,500 of them being microcalcifications randomly distributed) and five features describing each signal.

The second approach, based on the forward sequential search, reduced the relevant features to only three: absolute contrast, standard deviation of the gray level of the signal, and the third-order moment of the contour sequence [10]. Again, a transactional database was obtained, containing 10,000 signals including 2,500 microcalcifications randomly distributed, and three features describing each signal. For the third approach, using the GA, the original transactional database with all 47 features was used.

To test these two feature selection methods, simple feedforward ANNs with the corresponding number of inputs were trained and tested. The architecture of these ANNs consisted of five and three inputs respectively, $2n + 1$ neurons in the hidden layer (where n is the number of inputs), and one output. All the units had the hyperbolic tangent function as the transfer function. The data (inputs and targets) were scaled in the range [−1, 1] and divided into 10 non-overlapping splits, each one with 90% of the data for

Table 4.2.3 Results of the classification of individual microcalcifications

Method	Accuracy (%)	Sensitivity (%)	Specificity (%)
CS	84.56	50.91	95.94
FSS	81.33	76.21	81.92
GA	95.40	83.33	94.87

training and the remaining 10% for testing. Ten fold cross-validation trials were performed; that is, the ANNs were trained 10 times, each time using a different split on the data, and the averages of the overall performance, sensitivity, and specificity were reported. These results are shown in Table 4.2.3, representing the ANNs that had the best performance in terms of overall accuracy (percentage of correctly classified microcalcifications). The sensitivity (percentage of true positives or correctly classified microcalcifications) and specificity (percentage of true negatives or correctly classified objects that are not microcalcifications) of these ANNs are also shown.

For the third method, a GA was combined with ANNs to select the features to train them, as described earlier. The GA had a population of 50 individuals, each one with a length of $l = 47$ bits, representing the inclusion (or exclusion) of each one of the 47 features extracted from the signals. We used a simple GA, with Gray encoding, stochastic universal sampling selection, single-point crossover, fitness-based reinsertion, and a generational gap of 0.9. The probability of crossover was 0.7 and the probability of mutation was $1/l$, where l is the length of the chromosome (in this case, $1/l = 1/47 = 0.0213$). The initial population of the GA was always initialized uniformly at random. All the ANNs constructed by the GA are feedforward networks with one hidden layer. All neurons have biases with a constant input of 1.0. The ANNs are fully connected, and the transfer functions of every unit is the hyperbolic tangent function. The data (input and targets) were normalized to the interval $[-1, 1]$. For the targets, a value of "-1" means "non-microcalcification" and a value of "1" means "microcalcification". For training each ANN, backpropagation was used, only one split of the data was considered (90% for training and 10% for testing) and the training stopped after 20 epochs. The GA ran for 50 generations, and the results of this experiment are also shown in Table 4.2.3 in the row labeled "GA".

The best solution found with the GA method was an ANN with 23 inputs (five related to the contrast of the signal, four related to the background contrast, two related to the relative contrast, seven related to the shape, four moments of the contour sequence, and only one of the invariant geometric moments), corresponding to 48.94% of the original 47 extracted features. The ANNs coded in the chromosomes of the final population of the GA use 20.02 inputs on average, that is, the ANNs with the best performance need only 42.60% of the original 47 features extracted from the microcalcifications.

4.2.3.3 Microcalcification Clusters Detection and Classification

The process of cluster detection and the subsequent feature extraction phase generates another transactional database, this time containing the information on every microcalcification cluster detected in the images. A total of 40 clusters were detected in the 22 mammograms from the MIAS database that were used in this study. According to MIAS additional data and the advice of expert radiologists, 10 clusters are benign and 30 are malignant.

After the 30 features were computed from each microcalcification cluster, the first method for feature selection, based on class separability for ranking the features, reduced the relevant features to five: maximum radius, convex perimeter, standard deviation of the distances between microcalcifications, minimum absolute contrast, and standard deviation of the mean gray level of the microcalcifications in a cluster. The second approach, based on the forward sequential search, reduced the relevant features to only three: minimum diameter, minimum radius and mean radius, of the clusters.

The same procedure for the evaluation of the first two feature selection methods for the classification of individual microcalcifications was applied for these methods for the case of classification of clusters. The results in terms of the overall accuracy, sensitivity, and specificity of the best ANNs are shown in Table 4.2.4.

A GA method was also used to select the features for training ANNs, as described earlier. In this case, the transactional database containing the 30 features extracted from the clusters was used. The GA used in this case was similar to that used for the classification of signals, differing basically in the size of the chromosome and the mutation probability determined by the total number of features (30). For the targets, a value of "-1" means that the

Table 4.2.4 Results of the classification of microcalcification clusters

Method	Accuracy (%)	Sensitivity (%)	Specificity (%)
CS	84.56	50.91	95.94
FSS	77.5	53.85	88.890
GA	100.00	100.00	100.00

cluster is "benign" and a value of "1" means "malignant". The results of this experiment are also shown in Table 4.2.4 in the row labeled "GA".

The best ANN obtained with the GA method had nine inputs, corresponding to 30% of the original cluster feature set (five features related to the shape of the cluster, one related to the area of the microcalcifications, and three related to the contrast of the microcalcifications). On average, the chromosomes of the last generation coded 14.03 inputs; that is, the ANNs with the best performance only receive 46.76% of the original features extracted from the microcalcification clusters.

4.2.4 Conclusions and Future Work

We found that the use of GAs combined with ANNs greatly improves the overall accuracy, specificity, and sensitivity of the recognition of microcalcifications. We also found that all the ANNs coded in the chromosomes of the final population of the GA use 20.02 inputs on average; that is, the ANNs with the best performance need only 42.60% of the original 47 features. As an additional note, the first method, based on ranking features by class separability, had similar results in the case of the specificity, and a good overall performance, but had poor accuracy for the specificity of the classification.

In the case of the classification of microcalcification clusters, we also observed that the use of a GA for feature selection greatly improved the overall accuracy, sensitivity, and specificity, achieving values of 100%. On average, the best ANN architectures receive 14.03 inputs on average; that is, they only receive 46.76% of the 30 original cluster features as inputs. Nevertheless, only 40 microcalcification clusters were detected in the 22 mammograms used in this study. The test sets used in the ten fold cross-validation trial were very small and, in some splits, all the examples belonged to only one of the two classes, so either sensitivity or specificity could not be calculated. These splits were ignored in the calculation of the respective mean. On the other hand, the first two methods, based on ranking features by class separability and forward sequential search, had similar performance in terms of specificity and overall performance, but both showed deficient results for the specificity of the classification.

As future work, it would be useful to include and process other mammography databases, in order to have more examples and produce transactional feature databases that are more balanced and complete, and also test how different resolutions could affect system effectiveness. The size of the gaussian filters could be adapted depending on the size of the microcalcifications to be detected and the resolution of images. Different new features could be extracted from the microcalcifications in the images and tested, too. In this study, simple GAs and ANNs were used, and more sophisticated versions of these methods could produce better results. The inclusion of simple backpropagation training in the EANNs has consequences in terms of longer computation times, so alternatives to backpropagation should be tested in order to reduce time costs.

References

[1] Anttinen, I., Pamilo, M., Soiva, M., and Roiha, M.: Double reading of mammography screening films: one radiologist or two? *Clinical Radiology* **48** (1993) 414–421.

[2] Balakrishnan, K. and Honavar, V.: Evolutionary design of neural architectures. A preliminary taxonomy and guide to literature. Iowa State University, Department of Computer Sciences. Technical Report CS TR 95-01 (1995).

[3] Cantú-Paz, E. and Kamath, C.: Evolving neural networks for the classification of galaxies. In Proceedings of the Genetic and Evolutionary Computation Conference (GECCO 2002), San Francisco, CA, USA (2002), pp. 1019–1026.

[4] Cantú-Paz, E.: Feature subset selections, class separability and genetic algorithms. Lawrence Livermore National Laboratory, Center for Applied Scientific Computing (2004).

[5] Cantú-Paz, E., Newsam, S., and Kamath, C.: Feature selection in scientific applications. Lawrence Livermore National Laboratory, Center for Applied Scientific Computing (2004).

[6] Chandrasekhar, R. and Attikiouzel, Y.: Digitization regime as a cause for variation in algorithm performance across two mammogram databases. The University of Western Australia, Centre for Intelligent Information Processing Systems, Department of Electrical and Electronic Engineering. Technical Report 99/05 (1999).

[7] Dengler, J., Behrens, S. and Desaga, J.F.: Segmentation of microcalcifications in mammograms. *IEEE Transactions on Medical Imaging* **12**(4) (1993) 634–642.

[8] Ganott, M.A., Harris, K.M., Kaman, H.M., and Keeling, T.L.: Analysis of false-negative cancer cases identified with a mammography audit. *The Breast Journal* **5**(3) (1999) 166–175.

[9] Gulsrud, T.O.: Analysis of mammographic microcalcifications using a computationally efficient filter bank. Stavanger University College, Department of Electrical and Computer Engineering, Technical Report (2001).

[10] Gupta, L. and Srinath, M.D.: Contour sequence moments for the classification of closed planar shapes. *Pattern Recognition* **20**(3) (1987) 267–272.

[11] Hong, B.W. and Brady, M.: Segmentation of mammograms in topographic approach. IEE International Conference on Visual Information Engineering, Guildford, UK (2003).

[12] Kozlov, A. and Koller, D.: Nonuniform dynamic discretization in hybrid networks. In Proceedings of the 13th Annual Conference of Uncertainty in AI (UAI), Providence, RI, USA (2003), pp. 314–325.

[13] Kurkova, V.: Kolmogorov's theorem. In *The Handbook of Brain Theory and Neural Networks*. MIT Press, Cambridge, MA, USA (1995), pp. 501–502.

[14] Li, S., Hara, T., Hatanaka, Y., Fujita, H., Endo, T., and Iwase, T.: Performance evaluation of a CAD system for detecting masses on mammograms by using the MIAS database. *Medical Imaging and Information Science* **18**(3) (2001) 144–153.

[15] Ochoa, E.M.: Clustered microcalcification detection using optimized difference of gaussians. Air Force Institute of Technology, Wright-Patterson Air Force Base. Master Thesis (1996).

[16] Oporto-Díaz, S.: Automatic detection of microcalcification clusters in digital mammograms. Master Thesis, Center for Intelligent Systems, Tecnológico de Monterrey, Campus Monterrey, Monterrey, Mexico (2004).

[17] Oporto-Díaz, S., Hernández-Cisneros, R.R., and Terashima-Marín, H.: Detection of microcalcification clusters in mammograms using a difference of optimized gaussian filters. In Proceedings of the Second International Conference on Image Analysis and Recognition, ICIAR 2005, Toronto, ON, Canada (2005), pp. 998–1005.

[18] Suckling, J., Parker, J., Dance, D., *et al.*: The Mammographic Image Analysis Society digital mammogram database. *Exerpta Medica International, Congress Series* **1069** (1994) 375–378.

[19] Thurfjell, E.L., Lernevall, K.A., and Taube, A.A.S.: Benefit of independent double reading in a population-based mammography screening program. *Radiology* **191** (1994) 241–244.

[20] Yao, X.: Evolving artificial neural networks. *Proceedings of the IEEE* **87** (1999) 1423–1447.

4.3

Hybrid Detection of Features within the Retinal Fundus using a Genetic Algorithm

Vitoantonio Bevilacqua[1,2], Lucia Cariello[1,2], Simona Cambò[1], Domenico Daleno[1,2], and Giuseppe Mastronardi[1,2]

[1]*Department of Electrical and Electronics, Polytechnic of Bari, Bari, Italy*
[2]*e.B.I.S. s.r.l. (electronic Business In Security), Spin-Off of Polytechnic of Bari, Bari, Italy*

4.3.1 Introduction

The extensive use of information and communication technology in medicine is aimed at improving detection, diagnosis and treatment of diseases and disorders. Public administration places healthcare as one of the most important social issues and, thus, encourages the development of policies which make healthcare more easily accessible, improve patient care standards, reduce inefficiency, and remove barriers to screening and treatment which are cost effective.

Genetic and Evolutionary Computation: Medical Applications   Edited by Stephen L. Smith and Stefano Cagnoni
© 2011 John Wiley & Sons, Ltd

One useful tool which can satisfy, or at least aid all of the above, is the use of image manipulation. Medical imaging, for example, at molecular and cellular levels will allow a diagnosis to be made before symptoms actually appear, as well as individualizing genomic-based therapy with pinpoint accuracy. Surgery of the future will also be bloodless, painless, and less invasive. It will be powered by medical imaging systems that focus on the disease and unleash energy therapy to destroy the target whilst preserving healthy tissue. The objective of medical imaging is, in fact, the detection of illness at the stage at which it is most likely to respond well to therapy. In many cases, this also means it is less costly as it will lead to fewer complications and shorter hospital recovery time. Furthermore, advances in information technology and digital imaging will be able to offer clinicians new ways to capture and utilize 3-D real-time images which are transmitted directly from the patient.

One of the specialist areas in which the use of medical imaging is essential is that of ocular diagnosis. It is said that the retina 'is to the eye as film is to a camera' as it consists of multiple layers of sensory tissue and millions of photoreceptors whose function is to transform light rays into electrical impulses that reach the brain via the optic nerve – where they are perceived as images (see Figure 4.3.1(a)). Just beneath the retina there is the choroid, which is a set of blood vessels that supply oxygen and glucose to the retina (see Figure 4.3.1(b)).

The retina and its blood vessels are unique to each individual and are protected from variation caused by exposure to the environment by being located internally. Awareness of the uniqueness of the retinal vascular pattern dates back to 1935 when two ophthalmologists studying eye disease, Drs Carleton and Goldstein, made a startling discovery: each and every eye has its own totally unique pattern of blood vessels. They subsequently

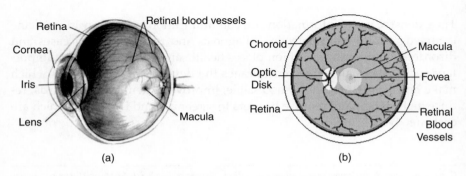

Figure 4.3.1 Side view (a) and rear view (b) of the eye

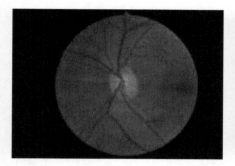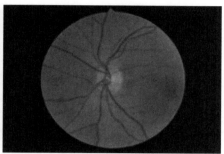

Figure 4.3.2 Twins' retinal fundi

published a paper on the use of retinal photographs for identification of people based on blood vessel patterns [1]. Later in the 1950s, their conclusions were supported by Dr Tower in the course of his study of identical twins [2]. He hypothesized that, when considering the variation of retinas in humans, identical twins would be most likely to have similar retinal vascular patterns (see Figure 4.3.2(a)).

However, Tower's study showed that, among the typical twin resemblance factors, retinal vascular patterns are those which have the least similarity. Moreover, among human physical features, none is longer lasting than the retinal fundus – it is stable from birth to death unless the person is afflicted by an eye disease.

Identification and study of retinal blood vessels can be a very useful process. Changes in retinal vasculature bifurcations and crossover points can be used for ocular screening purposes, the characterization of the history of retinal diseases or for the monitoring of treatment methods.

Alteration in the retinal vasculature can be tracked through observation of bifurcation and modification of crossover points, which may also provide clues regarding serious, progressive ocular illness. For example, glaucomatous changes are associated with a reduced number of nerve fibers, thus leading to changes in the optic nerve head configuration [3], and abnormal blood vessel growth and leaking blood vessels are causes of vision loss in eye conditions such as diabetic retinopathy, ROP, and macular degeneration (see Figure 4.3.3).

Manual measurements taken by a visual inspection are time-consuming, tedious or even impossible (e.g. when the vascular network is complex, or the signal-to-noise ratio is weak [4]). The development of an instrument to automate the analysis process, therefore, is important to overcome the

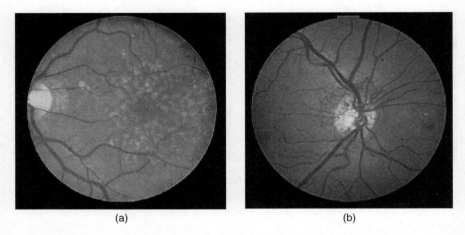

(a)   (b)

Figure 4.3.3 Patients' retinal images of (a) glaucoma; (b) diabetic retinopathy

disadvantages of visual analysis [5, 6]. Computer-aided diagnosis reduces the doctor's level of uncertainty regarding the diagnosis of some diseases. It improves precision of initial and evolutional disease identification, facilitates patient health status monitoring throughout new therapeutic treatment stages, stores images in digital format, and generates a diagnostic database that can be consulted and used in research, medical practice and specialized teaching. In addition, these diagnostic systems are potentially able to contribute to significant resource savings as well as being unaffected by observer bias and fatigue.

However, in order to provide an automatic support system for diagnosis, it is necessary to have an image acquisition/elaboration system to capture and process the retinal fundus image so that the physician is able to extract those features in which he is interested.

4.3.2 Acquisition and Processing of Retinal Fundus Images

Automatic acquisition and processing of images of the retinal fundus involves:

- acquiring images of the retinal fundus through an image capture device such as a camera;
- transferring the images to a processing system;
- automatically analyzing the acquired images.

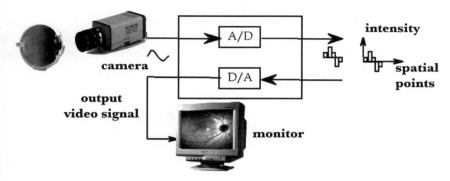

Figure 4.3.4   Digital processing system for analysis of retinal fundus images

The structure for the proposed digital processing system for images is represented schematically in Figure 4.3.4.

4.3.2.1 Retinal Image Acquisition

Acquisition of an image is achieved through the use of a camera or an optic sensor which generates analogue signals. The image signal is sampled by an analogue-to-digital converter and converted into a digital signal. Regardless of the field of application, a digital image must present the following characteristics in order to be processed effectively:

- acquisition of the image must be realized in optimum conditions – noise must be minimized or non-existent and repeatable measurements must be supported if required;
- the signal-to-noise ratio must be as high as possible;
- the image must be of sufficiently high resolution.

Digitalization transforms the continuous signal into quantized data both in terms of its spatial resolution (image sampled points) as well as its intensity resolution (gray levels). In order for the operator to perceive a continuous grayscale without quantization, it is usually necessary to use at least 8-bit resolution with 256 gray levels.

Photography of the ocular fundus can be achieved through either the use of fluorescein angiography, an ophthalmoscope or laser scans. They are all based on a similar principle in which the pupil is dilated and light reflected from the ocular fundus is recorded. This method is illustrated in Figure 4.3.5.

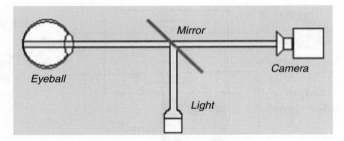

Figure 4.3.5 Simplified diagram of ocular fundus image observation and acquisition

Numerous acquisition systems have been designed involving the use of a computer and a camera interfaced to it. Differences between the various methods employed focus on the technology with which the camera has been built and its acceptability to the patient – an ever-important element to consider as retinal fundus acquisition is easily complicated by eye movement and the difficulty experienced by the patient when being measured. The use of contrast agents is also another differentiating factor. Fluorescein angiography is an example of a method which uses fluorescent sodium as a contrast agent which is injected into the bloodstream. Fluorescent sodium colorant (NaFl, $C_{20}H_{10}O_5Na_2$) is used because it is non-toxic and highly fluorescent [7], and quickly penetrates those areas where pathologies are present. The sodium is widespread throughout the human body within 10 to 20 minutes after injection, thus allowing retinal fundus screening to be carried out almost immediately.

4.3.2.2 Image Processing

Patient movement, poorly arranged imaging, bad positioning, reflections, opacity due to injury and inadequate lighting are all factors that can contribute to poor image acquisition, which affects subsequent analysis. Moreover, eye sphericity, a determinant factor in the intensity of retina tissue reflection, results in features such as low-frequency circular contrasts and intensity changes increasing at the interface between the front and rear eyepieces of the ocular camera. So, once the computer has acquired the image, a preprocessing phase will be necessary (through the use of appropriate software) in order to obtain higher image quality. In this way, it will be easier to process the image and, thus, extract features for diagnosis.

Image preprocessing can be used to:

(1) Improve the acquired image quality: correcting images, as far as possible, that exhibit noise, blurring or deformation.
(2) Analyze images: highlighting those details which are not clearly visible such as surface, shape factors, concavity and skeleton.

In either case, the fundamental processing element to be considered is the neighborhood windowing operation which is configured accordingly to achieve the required image processing operation.

Once this has been achieved the images are processed by applying the windowing operation to the entire image pixel-by-pixel. In this way, the intensity or gray-level matrix of the original image is transformed into the processed image. Many image processing and analysis algorithms have been developed to extract information from the photographed eye in order to help ophthalmologists in their diagnosis.

4.3.3 Previous Work

There have been many studies on the detection of blood vessels in medical images, but few of them are specifically related to retinal blood vessels. Studies on retinal image segmentation can be categorized into three main approaches. In the first category, studies based on line or edge detectors with boundary tracing can be grouped together. Wu *et al.* [8] developed a technique using computerized image analysis to measure the width of retinal arteries and veins on color fundus photographs. Width of the retinal vessel in color fundus photographs was determined by use of edge detection and boundary tracing-based programs.

A second category includes work based on matched filters, either 1-D profile matching with vessel tracking and local thresholding, or 2-D matched filters. Tolias and Panas [9] proposed an unsupervised fuzzy algorithm for vessel tracking to detect valid vessels in a fundus image. The method automatically tracks fundus vessels using linguistic descriptions like 'vessel' and 'non vessel' by the fuzzy C-means clustering algorithm. Additional procedures for checking the validity of detected vessels and the handling of junctions and forks are also implemented.

Zhou *et al.* [10] presented an algorithm that relies on a matched filtering approach, coupled with a priori knowledge about retinal vessel properties, to automatically detect vessel boundaries, track the vessel midline, and extract useful parameters of clinical interest. Gaussian functions were used to model the vessel profile and to estimate vessel diameter. In regions where vessels were straight, the authors used an adaptive densitometry tracking technique based on local neighborhood.

Gao *et al.* [11] applied strategies and algorithms for the measurement of vascular trees and included methods for locating the bifurcation center, detecting vessel branches, estimating vessel diameter, and calculating the angular geometry at a bifurcation point.

Jiang and Mojon [12] present a general framework of adaptive local thresholding based on a verification-based multi-threshold probing scheme. They allowed the classification procedure to accept part of a region by a series of different thresholds. By combining the results from the individual thresholds they obtained complete segmentation. The approach adopts the paradigm of hypotheses generation and verification; object hypotheses are generated by binarization using hypothetical thresholds which are accepted or rejected by a verification procedure. The application-dependent verification procedure can be designed to fully utilize all relevant information about the objects of interest.

Chaudhuri *et al.* [13] presented an algorithm that is based on directional 2-D matched filters in which an operator for feature extraction based on the optical and spatial properties of objects to be recognized is used. The gray-level profile of the cross-section of a blood vessel is approximated by a Gaussian-shaped curve and piecewise linear segments of blood vessels detected using the concept of matched filter detection of signals.

Zana and Klein [14] presented an algorithm based on mathematical morphology and linear processing for vessel recognition in a noisy retinal angiograph. Vessels are separated from their environment by a geometrical model of undesirable patterns: bright round peaks are first extracted, allowing segmentation of micro-aneurisms from images of diabetic patients; linear structures are then extracted using mathematical morphology and appropriate differential properties are computed using a Laplacian filter. The curvature differentiation is used to detect retinal vessels.

Lowell *et al.* [15] concentrated their efforts to realize a system to measure vessel diameter to sub-pixel accuracy based on fitting a local 2-D vessel model, which can measure vascular width to an accuracy of about one-third of a pixel. The diameter measurement is based on a two-dimensional difference of Gaussian model, which is optimized to fit a two-dimensional intensity vessel segment.

The third category regards those approaches that realize retinal blood vessel segmentation by supervised methods and require manually labeled images for training. Staal *et al.* [16] presented an algorithm based on extraction of image ridges, which coincide approximately with vessel centerlines. Ridges are used to compose primitives in the form of line elements to partition the images into patches. This is done by assigning each image pixel to the closest line element. Every line element constitutes a local coordinate frame for its corresponding patch. The feature vectors, for every pixel, are calculated by properties of the patches and the line elements, and are classified using a neural network classifier and sequential forward feature selection.

Iqbal *et al.* [17] identified bifurcation and retinal vasculature cross-over points by applying digital image processing, fuzzy logic and neural networks. At first, they used a preprocessing stage on the acquired images for illumination equalization and noise removal. Then, a fuzzy C-means algorithm, that clusters the image into two distinct classes, was used for segmentation. Detection of bifurcation and cross points was achieved using a 5×5 pixel neighborhood windowing operation, modified cross-point number method and neural network technique.

Bevilacqua *et al.* [18] presented a genetic algorithm-based approach for feature recognition using a database of several eye fundus images. In particular, the adoption of different genetic algorithms, each with its fitness function and chromosome structure was described for segmentation, vessel edge detection and bifurcation points identification in the processed retinal images. For each vessel's cross, the three or four bifurcation points of the detected edge were analyzed in order to achieve a more precise identification, then, for each cross, a unique bifurcation point was identified. For this reason, the authors chose the edge detection algorithm to avoid false edge points and to encourage narrow and continuous contours. Each phase was then formulated as an optimization problem with possible constraint handling, and several genetic algorithms, already implemented by the authors, were used to perform a parallel search in a space with a remarkable number of local minima where other techniques could fail.

4.3.4 Implementation

The following paragraphs describe the various steps of the technique developed by the authors to extract bifurcation and crossover points of retinal fundus images.

A database of 12 retinal fundus images was acquired using fluorescent angiography. Each image has a resolution of 512×512 pixels and a quantization

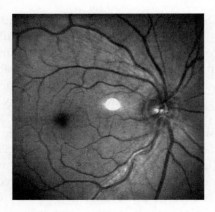

Figure 4.3.6 Example retinal fundus image

of 8 bits per pixel providing 256 gray levels. An example of an image used is depicted in Figure 4.3.6.

4.3.4.1 Vasculature Extraction

In the first step of the proposed process a combination of algorithms was used for the extraction of retinal features. At this stage, output is a precise representation of the vasculature without any alteration. It is important to note that the retinal fundus images are characterized by similar gray levels for both the background and vasculature features. This makes processing more challenging and is due to the presence of impulsive noise emitted by the acquisition instruments and because of structural noise emitted by the anatomic shape of the retinal fundus. The retina is, in fact, convex in shape, but depth information is not presented in the image.

For these reasons, and in order to characterize the required retinal features, a well-constructed image preprocessing phase was necessary and, hence, a system based on non-linear filtering power was used to render the retina image uniform. This required an operation that compacts gray levels effectively, as opposed to classic threshold systems which can lose large amounts of important information.

The filter chosen to meet our needs is widely known in literature as the Naka–Rushton filter [19], and is illustrated in equation (4.3.1).

$$O(i, j) = \frac{I(i, j)}{I(i, j) + \mu_{window}} \tag{4.3.1}$$

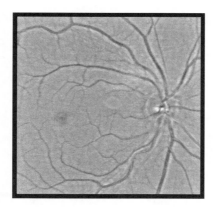

Figure 4.3.7   Naka–Rushton filtered image

where $O(I, j)$ is the output matrix that is the transformation result; $I(I, j)$ is the processed image matrix and μ_{window} is the mean average of pixels in the respective neighborhood processing window. This expression represents a compression filter and produces an equalization of the image gray levels, thus permitting greater contrast between the background and foreground objects contained within the image. The filter gives a simple and realistic vision of human color perception. A narrowing range in which levels of gray are distributed is achieved by this technique, which results in a uniform image, a feature that will facilitate the extraction of blood vessels. The result of this process is illustrated in Figure 4.3.7.

The filter action is an equalization of gray levels, as can easily be seen in the original and post-filter image histograms (see Figures 4.3.8 and 4.3.9, respectively). Considerable narrowing of the gray levels for the image histogram can be noted and, as a result, this permits the filter to be applied in order to separate blood vessels from the background image. Extraction of the vascular system image, however, is still difficult to implement given the presence of impulsive and structural noise. This cannot be removed without affecting other objects contained in the image which need preserving as they are of great importance to retain the exact form of objects.

At this point the application of a smoothing filter does little to achieve the desired result; these filters, in fact, only expand areas where there is noise. Application of a fixed image threshold results in loss of detail. Therefore, the post-preprocessing image was filtered in order to extract the vasculature through image clusterization [20]. The image was split into two clusters: vasculature and background clusters. This is an iterative method that begins with

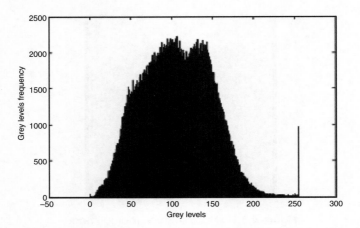

Figure 4.3.8   Original image histogram

the use of the average and standard deviation and then uses the Minkowski distance to distinguish vasculature pixels from their background. The output image has two levels: black and white pixels to represent the vasculature and background respectively (see Figure 4.3.10). The algorithm behind this filter incorporates relevant domain knowledge in order to achieve the robustness and feasibility required for the medical application. At this point, impulsive noise is still present in the image. Through the successive application of

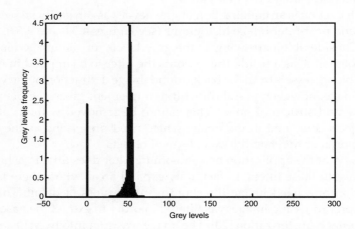

Figure 4.3.9   Retinal fundus image histogram after Naka–Rushton filtering

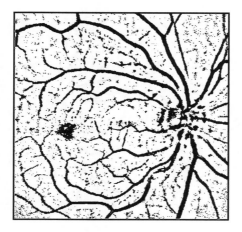

Figure 4.3.10 Clusterized image

morphological erosion operators (Figure 4.3.11), dilation operators (Figure 4.3.12) and median filtering (Figure 4.3.13), an image in which noise is reduced and features preserved is obtained. Specifically, erosion and dilation operations use a hyperbole filter [21–23], in the first case, on a negative image with a 17 × 17 pixel window and, in the second case, on an image with a 3 × 3 pixel window. The reason why the operator needs to be applied to the

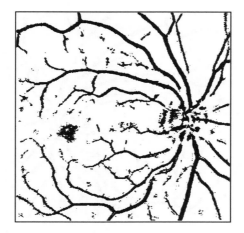

Figure 4.3.11 Image after application of morphological erosion operators

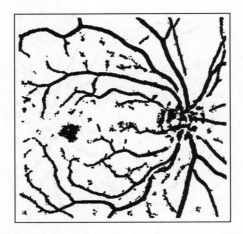

Figure 4.3.12   Image after application of morphological dilation operators

negative image lies in the fact that, in the first configuration, the filter was an erosion operator and that otherwise, the application of a white background to a black object renders the function similar to an expansion operator. As anticipated, at this stage it was appropriate to apply a further filter to normalize the image. The median filter, in addition to regulating the image, is also able to retain image characteristics without losing detail.

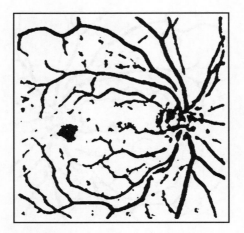

Figure 4.3.13   Image after median filtering

4.3.4.2 A Genetic Algorithm for Edge Extraction

The necessity to normalize the image as described in the previous paragraph derives from the fact that the genetic algorithm requires the image to be free from noise, and the objects of interest to be well-defined, in order to work effectively. The vessel extraction procedure has been realized by adapting a previously published edge detection method [18, 21]. It employs a genetic approach in which the problem is formulated as one of function cost minimization. In particular, it is necessary that the edge detection algorithm preserves properties of the vessel's edges. The semantic interpretation and recognition of the observed object has been founded on edge detection in which sequences of detached edges form closed lines without gaps or occlusions.

The search for an optimal solution to the objective function is performed through an iterative procedure applied to a population of chromosomes, corresponding to feasible solutions to the problem. The genetic algorithm implements a multi-directional search, maintaining a population of such potential solutions and encouraging the exchange of information between them through crossover operations. The cost function adopted is that proposed by Tan [24] for the measurement of edge fitness from real images. It introduces a cost factor related to the correct localization of edges based on a criterion of differences between adjacent regions.

The point cost of a binary edge image S at the position $p = (i, j)$ is a weighted sum of cost factors:

$$F(S, p) = \sum_i W_i C_i(S, p) \tag{4.3.2}$$

The total cost $F(S)$ of an edge image S is given by summing the point cost at each pixel of the image:

$$F(S) = \sum_p F(S, p) \tag{4.3.3}$$

Therefore, considering two edge images S_a and S_b identical apart from a 3×3 pixel region centered at position p, a comparative cost function can be defined as follows:

$$\Delta F(S_a, S_b, p) = \sum_{I(p)} \sum_k W_k [C_k(S_a, p) - C_k(S_b, p)] = \sum_{I(p)} \sum_k \Delta C_k(S_a, S_b, p) \tag{4.3.4}$$

where $0<=C_k<=1$ and $W_k>=0$. The C_k terms are the cost factors and the W_k terms the corresponding weights. In this arrangement, if $F(S_a, S_b, p) < 0$, S_a corresponds to a configuration of edges better than S_b, while it is the contrary when $F(S_a, S_b, p) > 0$. If $F(S_a, S_b, p) = 0$ the two configurations are equivalent in terms of cost. In relation to the cost C_t, the local attributes C_c and C_f mean the exclusive dependence of such factors on the configuration is assumed from the contours in the 3×3 pixel region observed, while C_d depends only on the value assumed in the observed position (i, j) within the image D of the dissimilarities. This is calculated in processing the dissimilarities image by means of template matching with suppression of non-maxima. The C_d cost is used to emphasize those pixels that constitute the common boundary between different adjacent regions and to introduce a penalty for those pixels not yet labeled as an edge. Therefore, the dissimilarities image, D, must have a real value at each p position in the interval $[0,1]$ (0 for no difference, 1 for maximum difference), proportional to the degree of dissimilarity between adjacent regions at that point. In order to derive this image, it is necessary first of all to determine a set of edge structures locally, valid for each element, forming two regions R_1 and R_2 on opposite sides of the structure on whose difference C_d will depend; therefore, it is necessary to determine a function $f(R_1, R_2)$ which computes the dissimilarities. In this case, 12 principal edge structures are locally valid, each consisting of two edge pixels, extended from the central pixel described in a 3×3 pixel window as shown in Figure 4.3.14.

For the sake of simplicity, the evaluation of the dissimilarity function $f(R_1, R_2)$ is based only on edges obtained from step edges in grayscale

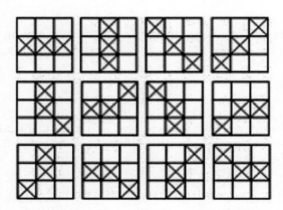

Figure 4.3.14 The 12 locally valid edge structures

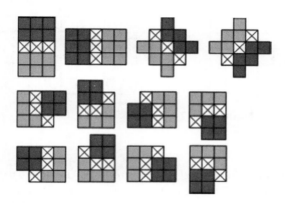

Figure 4.3.15   The regions to evaluate dissimilarities measure

signifying the boundary between the two regions of interest, as shown in Figure 4.3.15.

$$f(R_1, R_2) = \frac{1}{n_1} \sum_{p \in R_1} G(p) - \frac{1}{n_2} \sum_{p \in R_2} G(p) \qquad (4.3.5)$$

The cost, C_t, has been introduced to emphasize locally thin contours, applying a penalty for those that are thick. A pixel of a contour is considered when it exhibits multiple connections between two or more neighboring pixels within the 3×3 window area. The cost, C_t, of a point of an edge will be equal to 1 if considered thick according to the preceding definition, or equal to 0 if not (see Figure 4.3.16).

The C_c cost has been introduced to improve the structure of contours that are locally linear in spite of those locally curved. An edge pixel is considered to be locally linear, curved or very curved if the respective 3×3 surrounding window contains four linear structures, eight structures with 45° curves, or

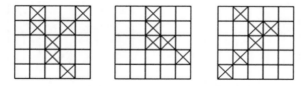

Figure 4.3.16 $C_t = 0$, $C_t = 1$, $C_t = 1$

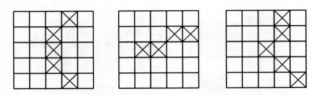

Figure 4.3.17 $C_c = 0$, $C_c = 0.5$, $C_c = 1$

any other structures. The C_c cost (see Figure 4.3.17) is set to 0 for a pixel which is not part of a curve, 0.5 for a pixel which is part of a curve, and 1 for a pixel which is part of a very curved line.

The C_f cost has been introduced to improve the structure of a contour that is locally continuous in spite of those bends. An edge pixel is considered to be linear, locally curved or very curved if the respective 3×3 surrounding window has more than one pixel of an adjacent contour, only one pixel of an adjacent contour, or no pixels of an adjacent contour. The C_f cost is therefore set to 0 for a linear pixel not bended, 0.5 for a curved pixel, and 1 for a very curved pixel (see Figure 4.3.18).

The C_e cost has been introduced to represent the number of pixels labeled like the contour, balancing the opposite inherent tendency in the C_d cost. C_e is set to 0 for a pixel not labeled like the contour, and to 1 for a pixel that is labeled like the contour. By using an optimal decision tree it is possible to improve the evaluation of the individuals' costs.

Concerning the weights attributed to the costs, Tan suggests a series of values that are generally valid, but are adapted on a heuristic basis with $C_e = 1.00$, $W_d = 2.00$, $W_c = \{0.25, 0.50, 0.75\}$, $W_f = \{2.00, 3.00, 4.00\}$; in case the local minima of the cost function corresponds to configurations of contours that are not thick, W_t is set equal to $W_f + W_d - W_c - W_e$.

Figure 4.3.18 $C_f = 0$, $C_f = 0.5$, $C_f = 1$

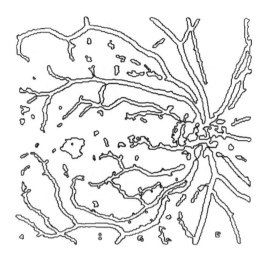

Figure 4.3.19   The final edge image

This algorithm, employing optimal code for the solutions and efficient mutation strategies, has produced the final edge image depicted in Figure 4.3.19.

4.3.4.3 Skeletonization Process

The next step consists of realizing a blood vessel skeleton from the template edges image obtained in the previous stage of processing. The process of skeletonization is divided into two steps: first of all, the edge image is processed along the horizontal direction and then along the vertical direction, always starting from the same image obtained in the previous processing step. Considering the image as a matrix of 512 rows and 512 columns, processing is applied to each row and column of the image using a horizontal scan and, subsequently, a vertical scan. Every pair of points that have a distance sufficient to be considered as a vessel edge are combined and then processed one by one to determine the median point. Each median value is then stored in the output image. Thus, we obtain two output images that contain the horizontal and vertical scan results and are used to calculate the final median point. The two images obtained are combined and the final image processed in order to eliminate the isolated points produced by the previous processing stage, which are not of any scientific relevance (see Figure 4.3.20).

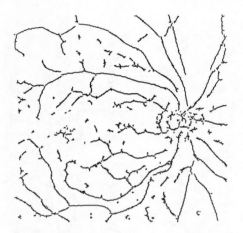

Figure 4.3.20   Complete skeleton image

4.3.4.4 Experimental Results

After obtaining the blood vessel skeleton, both the edge and skeleton images are merged in a unique view to exploit, simultaneously, the advantages of the information supplied from both sources (see Figure 4.3.21). At this point the application of a tracking algorithm allows the characterization of the bifurcation points.

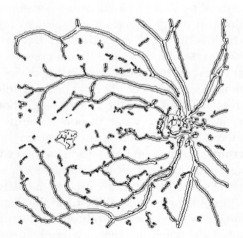

Figure 4.3.21   The edge image fused with the skeleton image

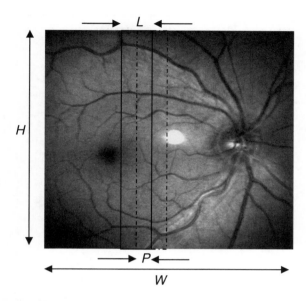

Figure 4.3.22 Retinal image parameterization and block extraction

In the tracking process, images of width 512 pixels (W in Figure 4.3.22) and height 512 pixels (H in Figure 4.3.22), are divided into overlapping blocks of height 32 (L in Figure 4.3.22) and width 512 (W in Figure 4.3.22). The amount of overlap between consecutives blocks is 16 pixels (P in Figure 4.3.22).

The idea of the algorithm is to start from the skeleton image and consider all pairs of points belonging to two parallel vessels that may ultimately converge towards a bifurcation point.

When a pair of vessels of this type are characterized, the tracking process starts with the pursuit of the points of the vessels contour under examination and, in particular, following the way covered by those edge points that are between the two skeleton points (see Figure 4.3.23).

Because of the different forms of blood vessels, it has been necessary to reinitiate the previous process, analyzing the images in all four directions: from left to right, right to left, top to bottom, and bottom to top, in order to cover all bifurcation and crossover points characterized by two vessels coming from different directions (see Figure 4.3.24).

The path of edge points is tracked by matching it, step by step, with preset schemas of routes that are different for each of the four scanning directions (see Figure 4.3.25).

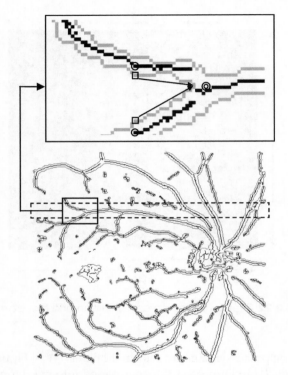

Figure 4.3.23   Tracking algorithm process

Where these tracks join, a bifurcation point is characterized. The obtained bifurcation and crossover points are depicted in Figure 4.3.26.

The test database of 12 images, containing 256 gray levels, was obtained from the oculist clinic of a local hospital. The present method has been applied to all images of the database, obtaining for each a bifurcation and crossover

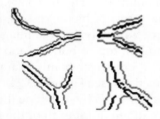

Figure 4.3.24   Examples of different forms of blood vessels in the four directions

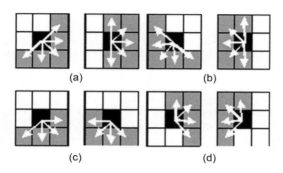

Figure 4.3.25 Possible routes of the two edge points considered in scanning (a) from left to right, (b) from right to left, (c) from top to bottom, (d) from bottom to top

point template that is an improvement over other methods tested. In the sequence of steps that lead to the result, an important role is played by the genetic algorithm that enables a faithful representation of retinal blood vessel edges to be realized. Consequently, the subsequent tracking algorithm is able to make a detailed analysis of bifurcations and crossover points in the examined image. In this way it is possible to detect a satisfactory number of searched points in a reasonably short time. The results obtained encourage the authors to continue the studies in this direction.

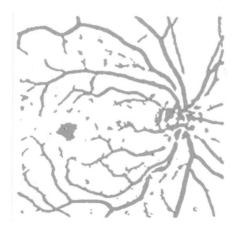

Figure 4.3.26 The obtained bifurcation and crossover points

References

[1] Simon, C., Goldstein, I.: A new scientific method of identification. *New York State Journal of Medicine*, Vol. 35, pp. 901–906 (1935).

[2] Tower, P.: The fundus oculi in monozygotic twins: report of six pairs of identical twins. *Archives of Ophthalmology*, Vol. 54, pp. 225–239 (1955).

[3] Paulus, D., Chastel, C., Feldmann, T.: Vessel segmentation in retinal image. *Medical Imaging*. SPIE Proceedings, Bd. Conference 5746 S., pp. 696–705 (2005).

[4] Zhang, Y., Hsu, W., Lee, M.: Segmentation of retinal vessels using nonlinear projections IEEE INternational Conference on IMage Processing, pp. 541–544 (2007).

[5] Goldbaum, M., Moezzi, S., Taylor, A., Chatterjee, S., Boyd, J., Hunter, H., Jain, R.: Automated diagnosis and image understanding with object extraction, object classification, and inferencing in retinal images. IEEE International Conference on Image Processing, pp. 695–698 (1996).

[6] Martinez-Perez, M.E., Hughes, A.D., Thom, S.A., Bharath, A.A., Parker, K.H.: Segmentation of blood vessels from red-free and fluorescein retinal images. *Medical Image Analysis*, Vol. 11, pp. 47–61 (2006).

[7] Bernardes, R., Dias, J., Cunha-Vaz, J.: Mapping the human blood–retinal barrier function. *IEEE Transactions on Biomedical Engineering*, Vol. 52, pp. 106–116 (2005).

[8] Wu, D.-C., Schwartz, B., Schwoerer, J., Banwatt, R.: Retinal blood vessel width measured on color fundus photographs by image analysis. *Acta Ophthalmology Scandinavia Suppl.*, Vol. 215, pp. 33–40 (1995).

[9] Tolias, Y., Panas, S.: A fuzzy vessel tracking algorithm for retinal images based on fuzzy clustering. *IEEE Transactions on Medical Imaging*, Vol. 17, pp. 263–273 (1998).

[10] Zhou, L., Rzeszotarski, M., Singerman, L., Chokreff, J.: The detection and quantification of retinopathy using digital angiograms. *IEEE Transactions on Medical Imaging*, Vol. 13, pp. 619–626 (1994).

[11] Gao, X., Bharath, A., Stanton, A., Hughes, A., Chapman, N., Thom, S.: Quantification and characterisation of arteries in retinal images. *Computer Methods in Programming Biometrics*, Vol. 63, pp. 133–146 (2000).

[12] Jiang, X., Mojon, D.: Adaptive local thresholding by verification based multithreshold probing with application to vessel detection in retinal images. *IEEE Transactions on Pattern Recognition and Analysis of Machine Intelligence*, Vol. 25, pp. 131–137 (2003).

[13] Chaudhuri, S., Chatterjee, S., Katz, N., Nelson, M., Goldbaum, M.: Detection of blood vessels in retinal images using two-dimensional matched filters. *IEEE Transactions on Medical Imaging*, Vol. 8, pp. 263–269 (1989).

[14] Zana, F., Klein, J.: Robust segmentation of vessels from retinal angiography. Proceedings of the International Conference on Digital Signal Processing, Santorini, Greece, pp. 1087–1090 (1997).

[15] Lowell, J., Hunter, A., Steel, D., Basu, A., Ryder, R., Kennedy, L.: Measurement of retinal vessel widths from fundus images based on 2-D modeling. *IEEE Transactions on Biomedical Engineering*, Vol. 23, pp. 1196–1204 (2004).

[16] Staal, J., Abramoff, M., Niemeijer, M., Viergever, M., van Ginneken, B.: Ridge-based vessel segmentation in color images of the retina. *IEEE Transactions on Medical Imaging*, Vol. 23, pp. 501–509 (2004).

[17] Iqbal, M.I., Aibinu, A.M., Nilsson, M., Tijani, I.B., Salami, M.J.E.: Detection of vascular intersection in retina fundus image using modified cross point number and neural network technique. Proceedings of the International Conference on Computer and Communication Engineering, pp. 241–246 (2008).

[18] Bevilacqua, V., Cariello, L., Introna, F., Mastronardi, G.: A genetic algorithm approach to detect eye fundus vessel bifurcation points. Proceedings of the International Conference on Computational Intelligence in Medicine and Healthcare, pp. 241–246 (2005).

[19] Naka, K.I., Rushton, W.A.: S-potentials from luminosity units in the retina of fish (*Cyprinidae*). *Journal of Physiology*, Vol. 185, pp. 587–599 (1966).

[20] Hsu, W., Pallawa, P.M.D.S., Lee, M.L., Eong, K.A.: The role of domain knowledge in the detection of retinal hard exudates. *CVPR IEEE*, Vol. 2, pp. 246–251 (2001).

[21] Bevilacqua, V., Mastronardi, G.: Edge detection using a steady state genetic algorithm. Proceedings of the 16th IMACS World Congress 2000 on Scientific Computation, Applied Mathematics and Simulation, pp. 215–8 (2000).

[22] Bevilacqua, V., Mastronardi, G.: Image segmentation using a genetic algorithm. Proceedings of the International Conference on Advances in Soft Computing. Springer-Verlag, pp. 111–123 (2002).

[23] Marino, F., Mastronardi, G.: Hy2: a hybrid segmentation method. Proceedings of the International Workshop on Image and Signal Processing, pp. 311–314 (1996).

[24] Tan, H. L., Gelfand, S. B., Delp, E. J.: A comparative cost function approach to edge detection. *IEEE Transactions on Systems, Man, and Cybernetics*, Vol. 19, pp. 1337–1349 (1989).

5

New Analysis of Medical Data Sets using GEC

5.1

Analysis and Classification of Mammography Reports using Maximum Variation Sampling

Robert M. Patton, Barbara G. Beckerman, and Thomas E. Potok

Oak Ridge National Laboratory, Oak Ridge, USA

5.1.1 Introduction

Currently, no automated means of detecting abnormal mammograms exist. While knowledge discovery capabilities through data mining and data analytics tools are widespread in many industries, the healthcare industry as a whole lags far behind. Providers are only just beginning to recognize the value of data mining as a tool to analyze patient care and clinical outcomes [8]. The research conducted by the authors investigates the use of genetic algorithms for classification of unstructured mammography reports, which can later be correlated to the images for extraction and testing.

In mammography, much effort has been expended to characterize findings in the radiology reports. Various computer-assisted technologies have been developed to assist radiologists in detecting cancer; however, the

Genetic and Evolutionary Computation: Medical Applications   Edited by Stephen L. Smith and Stefano Cagnoni
© 2011 John Wiley & Sons, Ltd

algorithms still lack high degrees of sensitivity and specificity, and must undergo machine learning against a training set with known pathologies in order to further refine the algorithms with higher validity of truth. In a large database of reports and corresponding images, automated tools are needed just to determine which data to include in the training set. Validation of these data is another issue. Radiologists disagree with each other over the characteristics and features of what constitutes a normal mammogram, and the terminology to use in the associated radiology report. Abnormal reports follow the lexicon established by the American College of Radiology Breast Imaging Reporting and Data System (Bi-RADS) [2], but even within these reports, there is a high degree of text variability and interpretation of semantics. The focus has been on classifying abnormal or suspicious reports, but even this process needs further layers of clustering and gradation, so that individual lesions can be more effectively classified [29]. The tools that are needed will not only help further identify problem areas, but also support risk assessment and other knowledge discovery applications.

The knowledge to be gained by extracting and integrating meaningful information from radiology reports will have far-reaching benefits, in terms of the refinement of the classifications of various findings within the reports. This will support validation, training, and optimization of these and other machine-learning and computer-aided diagnosis algorithms to work both in this environment and with other medical and imaging modalities. In the near-term, the objective of this work is to accurately identify abnormal radiology reports amid a massive collection of reports. The challenge in achieving this objective lies in the use of natural language to describe the patient's condition. The premise of this work is that abnormal radiology reports consist of words and phrases that are statistically rare or unusual. If this is true, then it is expected that abnormal reports will be significantly dissimilar in comparison to normal radiology reports.

To achieve this objective, our approach employs maximum variation sampling (MVS), which is implemented as an adaptive sampling approach [16, 32, 33]. Maximum variation sampling seeks to identify a particular sample of data that will represent the diverse data points in a data set. Adaptive sampling continues to draw samples from the population based on previous samples until some criteria have been met. Previous results from using MVS indicated that an ideal sample could be found very quickly using this approach [17, 18].

5.1.2 Background

Mammography is the procedure of using low-dose X-rays to examine the human breast for the purposes of identifying breast cancer or other abnormalities. Currently, for each patient that undergoes a mammogram, there is at least one X-ray image and one textual report written by a radiologist. In the report, the radiologist describes the features or structures that they see or do not see in the image. If an abnormality or suspicious area is found, the patient may undergo a diagnostic mammogram or biopsy, which results in additional images and reports in the patient's record. Essentially, these reports are meta-data about the corresponding image that is written by a human subject matter expert. In order to effectively train a computer-assisted detection (CAD) system, these reports could be mined and used as supplemental meta-data. Unfortunately, little work has been done to utilize and maximize the knowledge potential that exists in these reports.

There are several challenges in utilizing these reports. First, the reports vary in length. Some radiologists use more words than others when describing the same features. For example, in patients that do not exhibit any suspicious features, there are some reports that very simply state that there are no suspicious features. However, for the exact same patient with no suspicious features in a different year, a different radiologist will provide a much more lengthy report that describes all of the suspicious features that did not exist.

To provide a better perspective of the challenge of mining these reports, consider the following question. Given a database of these reports, how does one classify those reports that represent abnormalities in the patient? In mammography, most patient reports will represent "normal" conditions in the patient. Consequently, the reports with "abnormal" conditions are rare (defining the difference between what is "normal" and "abnormal" is beyond the scope of this work). Performing a cluster of these reports, most of the normal reports would cluster together while the abnormal reports would not form a cluster. This is because "abnormal" conditions tend to be unique and very specific to a patient, while "normal" conditions are much more generic and broad. Even if clustering provided value, clustering a very large database of these reports is exceptionally computationally expensive. Categorizing would be faster; however, the challenge remains of determining the appropriate categories, and even then, the abnormal reports may not categorize correctly.

Another challenge to utilizing these reports lies in the language that is used in mammograms. Abnormal reports tend to have a richer vocabulary than normal reports. In addition, normal reports tend to have a higher number of "negation" phrases. These are phrases that begin with the word "no", such as in the phrase "no findings suggestive of malignancy." Consider the phrases shown in Tables 5.1.1 and 5.1.2. These are the negation phrases that generally occur in normal reports and the ones shown here are samples of the variations that have been found. In the set of reports used for this work, there were at

Table 5.1.1   Example phrases using "no" and "malignancy"

no malignancy
no mammographic features malignancy
no mammographic features suggestive of malignancy
no findings suggestive of malignancy
no significant radiographic features of malignancy
no radiographic findings suggestive of malignancy
no radiographic change suggestive of malignancy
no specific radiograpic features of malignancy
no mamographic evidence of malignancy

least 286 variations of phrases for Table 5.1.1 and 1,231 variations of phrases for Table 5.1.2.

Consider the phrases shown in Tables 5.1.3 and 5.1.4. These phrases tend to occur in abnormal reports (but may also occur in normal reports) and the ones shown here are samples of the variations that have been found. In the set of reports used for this work, there were at least 52 variations of phrases for Table 5.1.3 and 691 variations of phrases for Table 5.1.4.

Considering the language variations shown previously, the task of classifying those reports that represent abnormalities is daunting. The variations of terms and syntax create a combinatorial explosion while, semantically, these combinations tend to mean the same thing.

5.1.3 Related Works

There has been considerable work in a variety of areas in the text analysis community and a wide array of problems with processing and analyzing text

Table 5.1.2   Example phrases using "no" and "suspicious"

no mammographic finding suspicious
no strongly suspicious forms
no strongly suspicious features
no strongly suspicious masses
no radiographically suspicious masses
no developing suspicious clustered microcalcifications
no finding strongly suspicious
no new suspicious mass lesions
no suspicious linear branching forms

Table 5.1.3   Example phrases using "appearance" and "tissue"

appearance suggesting radiating strands of tissue
appearance suggestive of accessory breast tissue
appearance of normal glandular tissue
appearance of asymmetric fibroglandular tissue
appearance of fibroglandular tissue
appearance of glandular tissue
appearance of normal fibroglandular tissue
appearance of soft tissue densities bilaterally

data. Some of these areas of text analysis include retrieval, categorization, clustering, syntactic and semantic analysis, duplicate detection and removal, and information extraction to name a few [4, 23, 24, 28, 34]. These areas range from analyzing entire data sets to analyzing a single document. In general, as the size of the data set increases, many of these approaches begin performing poorly, or the value of their results begins to diminish. For example, clustering usually requires comparing every document with every other document. Obviously, as the data set size increases, performance will noticeably suffer. However, with categorization, the performance may not suffer considerably, but the quality of the results will be diminished if a sufficient number of categories are not identified or if the categories are not clearly or accurately identified [27].

Further improvement in information retrieval techniques requires the continued development of algorithms whose basis lies in semantic extraction and representation. Information retrieval (IR) research began with simple representations of documents and the terms that they contained [26]. This research progressed into syntactic analysis such as co-occurrence, N-grams, part-of-speech analysis, and context-free grammars. Recently, IR research has

Table 5.1.4   Example phrases using "additional" and "views"

additional views obtained today demonstrate variation
additional compression views
additional set of bilateral cc views
additional lateral views
additional mediolateral oblique views
additional mammographic views
additional bilateral craniocaudal views
additional bilateral lateral medial views

continued to move toward a basis in semantics. Many of these approaches involve the use of ontologies, conceptual graphs, and language models such as described in [6, 7, 11, 12, 26]. Unfortunately, many of these approaches are either unable to scale, require significant effort on the part of subject matter experts, or do not handle domain-specific data robustly. The work described here differs from these approaches in that it leverages computationally efficient, unsupervised learning of domain-specific data in order to more effectively retrieve information. As a result, extensive ontologies are not needed, or extensive effort on the part of a subject matter expert.

In [1], an unsupervised approach to identifying cue phrases is discussed. Cue phrases are formulaic patterns of phrases that have similar semantics but vary in syntactical and lexical ways. In [1], the authors use a lexical bootstrapping algorithm that relies on the use of "seed" phrases. While our work is addressing nearly the same problem, our work differs in that no seed phrases are needed, and s-grams found for cue phrases using our approach are split into two classes.

Other work is being done in the medical environment to use automated software tools to extract knowledge from unstructured radiology reports [5]. Preliminary findings demonstrate that automated tools can be used to validate clinically important findings and recommendations for subsequent action from unstructured radiology reports. Commercially available software is also being tested to automate a method for the categorization of narrative text radiology reports, in this case dealing with the spine and extremities [31].

5.1.4 Maximum Variation Sampling

The objective of this work is to accurately identify abnormal radiology reports amid a massive collection of reports. As discussed earlier, abnormal reports have wider variation in their language than normal reports. Consequently, what is needed is to sample the most diverse reports and identify the common language that is unique to those reports. This common language will then provide the basis for classification.

Sampling can be divided into two main categories: probability-based and nonprobability-based. Probability-based sampling is based on probability theory and the random selection of data points from the data set. Nonprobability-based sampling is based on purposeful selection, rather than random selection. The advantage of this form of sampling is that it allows the analyst to look at data that may not otherwise be visible via the random selection process. In the domain of mammography reports, random selection would not easily find abnormal reports, as they constitute a very small portion of all reports.

Within nonprobability-based sampling, there are several categories of sampling [16], one of which is maximum variation sampling (MVS) [16]. This particular sampling method seeks to identify a particular sample of data that will represent the diverse data points in a data set. According to Patton [16], "This strategy for purposeful sampling aims at capturing and describing the central themes or principle outcomes that cut across a great deal of [data] variation." In a large text corpus, this form of sampling provides the ability to quickly characterize the different topics, or "threads" of information that are available.

A genetic algorithm (GA) was developed to implement the maximum variation sampling technique. Genetic algorithms are nature-inspired algorithms that mimic the natural selection process [10]. The natural selection process is generically defined as survival of the fittest (i.e., only the most fit individuals for a given environment survive and reproduce offspring). During this process, the offspring are created from the best individuals; therefore, the population should continue to improve over several generations. It has been shown that canonical genetic algorithms converge to an optimal solution if the best individual remains in the population [25].

It is well known that a genetic algorithm performs very well for large search spaces and is easily scalable to the size of the data set. In addition, GAs are also particularly suited for parallelization [13, 15, 30]. To better understand the need for scalability and the size of the search space in this problem domain, consider a set of 10,000 radiology reports. Now, suppose an analyst needs to reduce this data set to 200 representative reports (only 2% of the entire data set). In that case, there are approximately 1.7×10^{424} different combinations of reports that could be used to create a single sample. Clearly, a brute force approach is unacceptable. In addition, many of the combinations would consist of duplicate data that would lower the quality of the result for the analysts. Ultimately, an intelligent and scalable approach such as a genetic algorithm is needed. As demonstrated by Mutalik [14], a parallel genetic algorithm is well suited to a combinatorial optimization problem.

Before applying a GA to the analysis of radiology reports, the reports must be prepared using standard information retrieval techniques. First, reports are processed by removing stop words and applying the Porter stemming algorithm [9, 20, 21]. Once this has been done, the articles are then transformed into a vector-space model (VSM) [22, 26]. In a VSM, a frequency vector of word occurrences within each report can represent each report. Once vector-space models have been created, the GA can then be applied.

Sample Size is N

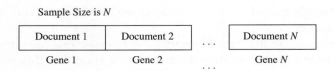

Figure 5.1.1 Genetic representation of each individual

Two of the most critical components of implementing a GA are the encoding of the problem domain into the GA population and the fitness function to be used for evaluating individuals in the population. To encode the data for this particular problem domain, each individual in the population represents one sample of size N. Each individual consists of N genes, where each gene represents one radiology report (each report is given a unique numeric identifier) in the sample. For example, if the sample size were 10, each individual would represent one possible sample and consist of 10 genes that represent 10 different reports. This representation is shown in Figure 5.1.1.

The fitness function evaluates each individual according to some predefined set of constraints or goals. In this particular application, the goal for the fitness function was to achieve a sample that represents the maximum variation of the data set without applying clustering techniques or without prior knowledge of the population categories. To measure the variation (or diversity) of our samples, the summation of the similarity between the vector-space models of each document (or gene) in the sample is calculated, as shown in the following equation:

$$Fitness(i) = \sum_{j=0}^{N} \sum_{k=j+1}^{N} \alpha_j + \beta_k + Similarity\ (Gene\ (i, j),\ Gene\ (i, k)) \qquad (5.1.1)$$

In Equation (5.1.1), the *Similarity* function calculates the distance between the vector space models of gene j and k of the individual i. This distance value ranges between 0 and 1, with 1 indicating that the two reports are identical and 0 indicating that they are completely different in terms of the words used in that report. Therefore, in order to find a sample with the maximum variation, Equation (5.1.1) must be minimized (i.e., lower fitness values are better). In this fitness function, there will be $(N^2 - N)/2$ comparisons for each sample to be evaluated.

In an effort to effectively characterize the phrase patterns of the mammography reports, it is necessary to examine reports that are longer in length, so that more language can be examined for patterns. The data set for this work contains numerous reports that simply state that the patient canceled their appointment. These reports are very short in length and are exceptionally distinct from all other reports (similarity values approaching zero). In addition, abnormal reports tend to be longer in length than normal reports since the radiologist is describing the abnormalities in more detail. Consequently, the fitness function of the MVS-GA incorporates penalty functions as shown in Equations (5.1.2) and (5.1.3):

$$\alpha_j = e^{-\left(\frac{\|j\|}{100}\right)} \tag{5.1.2}$$

$$\beta_k = e^{-\left(\frac{\|k\|}{100}\right)} \tag{5.1.3}$$

In the penalty equations, shorter documents receive higher penalties while longer documents receive much lower penalties. The penalty functions also return values that are between 0 and 1, inclusive. As a result of the penalty functions, the cancellation reports will receive the highest fitness values, while lengthy, abnormal reports will receive the lowest fitness values.

To create children from a given population, genetic operators such as selection, crossover, and mutation are applied to the individuals. For each generation, an average fitness value is calculated for the population. Individuals with fitness values that are above this average are selected as parents, while the other individuals are discarded. This can be a very aggressive selection process if there are extremely fit individuals that are far above the average. Once parents are selected, crossover and mutation operators are applied to the parents to create children. The crossover and mutation operators are 1-point operators [10]. After the MVS-GA is executed, the end result is a best sample of mammography reports that are as diverse from each other as possible.

In addition to finding a sample of the most diverse reports, the MVS-GA was also enhanced to extract the common skip bigrams (s-grams) from the reports. S-grams are word pairs in their respective sentence order that allow for arbitrary gaps between the words [1, 3, 19]. The s-grams for Table 5.1.1 are the words "no" and "malignancy." This s-gram uniquely identifies a particular semantic in the language of mammography reports and enables the identification of all possible variations of such phrases. Higher-level patterns may then be formed from these s-grams. For example, the s-grams for Tables 5.1.1 and 5.1.2 both imply that there are no abnormalities seen in the patient.

Once the best sample is achieved by the MVS-GA, then phrases are extracted from each document in the sample. For each phrase in the document, s-grams are extracted. Next, the s-grams are counted across the sample of documents. S-grams that are common across the sample will have higher frequency counts, while s-grams with a frequency of 1 uniquely identify a particular document in the sample. For this work, only those s-grams that are the most frequent in the best sample found are considered valuable. It is these s-grams that have the ability to uniquely classify abnormal documents from a large set.

The primary intent of the GA is to converge toward an optimal solution. However, very little GA research, if any, has been performed that leverages knowledge gained from the individuals that failed to be selected and reproduce. In a typical GA, individuals that are not selected for reproduction are simply discarded. For the MVS-GA, this is different. In addition to extracting s-grams from successful individuals, the MVS-GA has also been augmented to store the most frequent s-grams of the failed individuals. This will enable answering questions such as what characteristic phrases make failed individuals inferior to successful individuals. After each generation, s-grams and their frequencies from each failed individual are extracted from each individual and stored in memory. After the MVS-GA has completed, the memory now contains the most frequent s-grams that caused individuals to fail in the GA. Individuals that fail in the MVS-GA tend to contain a high number of normal reports. Successful individuals tend to contain a high number of abnormal reports. The end result is that the MVS-GA learns the most frequent s-grams for both abnormal and normal classes of reports.

5.1.5 Data

In this work, unstructured mammography reports were used. Each report generally consists of two sections. The first section describes what features the radiologist does or does not observe in the image. The second section provides the radiologist's formal medical opinion as to whether or not there are suspicious features that may suggest malignancy (i.e., the possibility that the patient has cancer).

These reports represented 12,809 patients studied over a 5-year period from 1988 to 1993. There are 61,064 actual reports in this set, which include a number of reports that simply state that the patient canceled their appointment. Table 5.1.5 and Figure 5.1.2 show the general statistics and distribution of the number of reports per patient.

According to Table 5.1.5, the positive skewness indicates that there are many patients with more reports than the average. Since the study was five

Table 5.1.5 Statistics of number of reports per patient

Minimum	1
Maximum	32
Average	4.77
Std Dev	3.57
Skewness	1.27
Kurtosis	1.76

years in duration, the vast majority of the patients will have either one report every year or one report every two years. This explains the average of 4.77 and kurtosis of 1.76. For some of the patients, abnormalities were identified that required additional diagnostic screenings. In addition, some patients have reports that predate the beginning of the study. In the extreme case of 32 reports, a patient record contained reports that predated the study by nearly 10 years and the patient also had breast cancer in the right breast and an abnormality in the left breast that was later determined to be benign. Consequently, these patients with the extra reports explain the positive skewness.

Furthermore, of this large data set, a human expert manually classified 100 reports as being normal and 100 reports as being abnormal. From the normal set of reports, a random sample of 90 reports was selected. From the abnormal set, a random sample of 10 reports was selected. The two samples were then merged to create a third set of 100 reports. This third set was used to initially

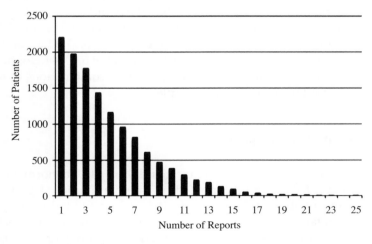

Figure 5.1.2 Distribution of the number of reports per patient

test the MVS-GA. If the premise that abnormal radiology reports consist of words and phrases that are statistically rare or unusual, then the expected result of the MVS-GA will be a sample of reports consisting predominantly of abnormal reports.

5.1.6 Tests

Two sets of tests were performed. For the first set of tests, the test data consisted of just 100 reports from the human classified group, where 90 of these reports were classified as normal and 10 were classified as abnormal. The objective of the first set of tests was to determine if, in fact, the MVS-GA could identify abnormal reports amid a significantly larger number of normal reports. After the first tests were performed, the MVS-GA was enhanced as described earlier to extract s-grams from the successful and failed individuals in the population. The objective of the second set of tests was to learn the key s-grams that could be used to classify normal and abnormal reports.

For the initial tests, 30 runs of the MVS-GA were performed. The population size was defined as 2,000 and the number of generations was set to 250. The crossover rate was set to 0.7 and the mutation rate to 0.03. The sample size was set to 15 and the data set size was 100 radiology reports (90 normal and 10 abnormal). In this case, there are approximately 2.53×10^{17} different combinations of reports that could be used to create a single sample.

For the second set of tests, the MVS-GA was enhanced to extract s-grams from both successful and failed individuals in the population. The population size was defined as 2,000 and the number of generations was set to 2,500. The crossover rate was set to 0.7 and the mutation rate to 0.03. The sample size was set to 100 and the data set size was the entire set of 61,064 reports. In this case, there are approximately 3.74×10^{320} different combinations of reports that could be used to create a single sample.

5.1.7 Results & Discussion

For the initial tests, the GA consistently found 8 out of 10 abnormal reports. The remainder of the sample consisted of 7 normal reports. Upon further analysis of the 10 abnormal documents, it was found that 4 of the reports were very similar to each other, while the other 6 were very distinct from each other. Consequently, 2 of the 10 abnormal reports were consistently absent from the final sample.

Upon further analysis of the normal documents that were included in the final sample, it was determined that several of the reports represented "boundary" cases. These were reports that, while considered normal by a human expert, represented situations where a patient had either already undergone a lumpectomy or had a family history of breast cancer and showed high potential for breast cancer. Other normal reports that were in the final sample consisted of patients who needed further examination and therefore underwent spot magnification for further confirmation. Another report represented a patient where the radiologist had difficulty in determining a nodule in the image and suggested that it was a "small deformable cyst." Overall, these preliminary results from the GA showed encouraging performance to find both abnormal reports and potentially unusual normal reports without prior categorization or a predefined vocabulary of terms to search.

Additional analysis of the final sample revealed another characteristic of the reports. For each report, word phrases unique to that specific report were extracted. In this case, unique word phrases are those phrases that only appear in one report in the sample. As shown in Table 5.1.6, normal reports tended to have fewer unique word phrases compared with abnormal reports. In addition, abnormal reports tended to have more variability in the number of unique word phrases, as shown by the standard deviation that is nearly twice that of the normal reports.

Further investigation into the word phrases of the abnormal reports revealed a wide-ranging vocabulary and semantics. Table 5.1.7 shows example word phrases from both normal and abnormal reports.

Table 5.1.6 Number of unique phrases for each report

Normal Reports	Abnormal Reports
18	26
15	63
11	38
14	43
16	29
0	45
23	22
–	27
Avg: 13.857	**Avg: 36.625**
Std Dev: 7.151	**Std Dev: 13.553**

Table 5.1.7　Sample word phrases from reports

Normal Reports	Abnormal Reports
benign biopsy	intraductal carcinoma
breasts unchanged	rod shaped calcifications
microcalcifications identified	defined hyperdense nodule
remain unchanged	hypoechoic lesion
small deformable cyst	recommend excisional biopsy
benign macrocalcification	lobulated hypoechoic mass

Analysis of the word phrases provides further evidence to support our hypothesis that abnormal reports consist of statistically rare or unusual words, thereby making them easier to identify in a large collection of reports.

After this first test, the MVS-GA was enhanced to extract s-grams from the individuals in the population. The s-grams discovered by the MVS-GA on the entire data set are shown in Tables 5.1.8 and 5.1.9. Table 5.1.8 shows the top 10 most frequently occurring s-grams from the best solution obtained

Table 5.1.8　Top 10 most frequently occurring s-grams from best solution obtained by MVS-GA

Rank	S-gram	Example Phrase	Number of Variations Observed
1	magnification & views	magnification views requested	660
2	core & biopsy	stereotactic guided core biopsy of microcalcifications	633
3	needle & localization	ultrasound-guided needle localization procedure	245
4	nodular & density	showing questionable increased nodular density	2726
5	lymph & node	atypically located intramammary lymph node	748
6	needle & procedure	stereotactic needle core biopsy procedure	57
7	compression & views	right anterior compression views	772
8	spot & views	recommended utilizing spot views	852
9	spot & compression	spot compression image	1123
10	spot & magnification	medially exaggerated right cc spot magnification	650

Table 5.1.9 Top 10 most frequently occurring s-grams with the word "no"

Rank	S-gram	Example Phrase	Number of Variations Observed
1	no & suspicious	no finding strongly suspicious	1225
2	no & calcifications	no clear-cut clustered punctate calcifications	137
3	no & evident	no mass lesions evident	46
4	no & masses	no new focal masses	365
5	no & malignancy	no specific evidence of malignancy	286
6	no & residual	no residual microcalcifications	56
7	no & skin	no skin abnormalities noted	68
8	no & thickening	no skin thickening seen	42
9	no & complications	no apparent complications	16
10	no & change	no apparent interval change	384

by the MVS-GA. These s-grams tend to uniquely define abnormal reports. Many of these s-grams refer to procedures that are performed in the event that a suspicious feature in the patient was observed by the radiologist. For example, the patient may be asked to return within a few weeks for additional imaging such as an ultrasound and magnification imaging. In addition, patients with suspicious features may undergo biopsy and, in some cases, may also have a needle localization performed to enhance the biopsy procedure. Furthermore, since breast cancer often affects the lymph nodes, radiologists look for abnormalities relating to the lymph nodes as well. As can be seen in Table 5.1.8, the MVS-GA successfully learned key s-grams that would significantly enhance automated classification of abnormal reports.

Table 5.1.9 shows the 10 most frequently occurring s-grams that begin with "no" and were learned from the failed individuals (i.e., individuals that were not selected for reproduction) in the MVS-GA. As discussed previously, most normal reports contain some form of a "negation" phrase. These phrases refer to the non-existence of a particular feature or condition in which the radiologist was searching. Abnormal reports may contain such negation phrases; however, abnormal reports tend to be more focused on the abnormalities that were found and not the abnormalities that were not found. Consequently, MVS-GA successfully learned from the failed samples the key s-grams of normal reports.

Table 5.1.10 Abnormal and normal class separability using most frequent extracted s-grams

	Using all Terms	Using only Abnormal and Normal S-Grams	Percent Change
Average "Normal" feature vector length	57.79	5.81	−89.94%
Average "Abnormal" feature vector length	117.79	9.57	−91.88%
Within-class similarity "Normal"	0.1118/0.1114 (avg/std dev)	0.8669/0.1070 (avg/std dev)	675.40%
Within-class similarity "Abnormal"	0.1306/0.1056 (avg/std dev)	0.3621/0.2999 (avg/std dev)	177.26%
Between-class similarity	0.0618/0.0876 (avg/std dev)	0.0531/0.1544 (avg/std dev)	−14.08%

After the most frequent s-grams were extracted for normal and abnormal reports using the MVS-GA, these s-grams were then analyzed for their ability to distinguish between the two classes of reports. For this analysis the data set used consisted of 100 reports classified as "Normal" and 100 reports classified as "Abnormal" by a human expert. The average similarity (using the cosine similarity measure for all similarity calculations) between the reports within the Normal class was computed. This same average was also computed for reports within the Abnormal class. Both of these averages are referred to as "Within-class Similarity." Next, the average similarity of reports between the two classes was computed and referred to as "Between-class Similarity." An ideal classifier should have a within-class similarity near 1.0 for each class and a between-class similarity near 0.0 for each pair of classes. Such a classifier would very accurately distinguish reports as being of a particular class.

Table 5.1.10 shows the results of using the most frequent normal and abnormal s-grams from Tables 5.1.8 and 5.1.9 to represent the report content. As can be seen, the ability to distinguish the normal and abnormal reports is very low when using all of the terms in the reports. This is due to the amount of noise in the language of the reports. However, when using only the normal and abnormal s-grams that exist in the reports, the improvement is considerable, especially for the normal class. Again, this is due to the fact that normal reports are shorter and less variable in their language, while abnormal reports are longer and more ambiguous in their language. Also notable is that this improvement in the separability between normal and abnormal reports was achieved using considerably fewer terms, as can be seen by the average feature vector lengths. Approximately 90% of the terms could be removed from each report while providing significant improvements in the within-class and between-class similarities.

5.1.8 Summary

Currently, text analysis of mammography reports remains a significant challenge. However, solving this issue would provide numerous benefits. The work described here represents results in applying a GA to assist with identifying abnormal mammography reports from a large set of reports. Results were very encouraging and show tremendous potential for future work. Future work will seek to leverage this technique to develop a more advanced and specific training set of images to further enhance image-based algorithms.

Acknowledgments

Our thanks to Robert M. Nishikawa, PhD, Department of Radiology, University of Chicago for providing the large data set of unstructured mammography reports, from which the test subset was chosen.

References

[1] Abdalla, R.M. and Teufel, S. A bootstrapping approach to unsupervised detection of cue phrase variants. *Proceedings of the 21st International Conference on Computational Linguistics and the 44th Annual Meeting of the Association for Computational Linguistics* (Sydney, Australia). COLING 2006. ACM Press, New York, 2006.

[2] American College of Radiology (ACR). ACR BI-RADS® – Mammography, 4th edn. *ACR Breast Imaging Reporting and Data System, Breast Imaging Atlas*. American College of Radiology, Reston, VA, 2003.

[3] Cheng, W., Greaves, C. and Warren, M. From n-gram to skipgram to concgram. *International Journal of Corpus Linguistics* **11**(4) (2006) 411–433.

[4] Cui, X. and Potok, T.E. A distributed agent implementation of multiple species Flocking model for document partitioning clustering. *Lecture Notes in Computer Science*, Vol. 4149. Springer-Verlag, Heidelberg, 2006, pp. 124–137.

[5] Dreyer, K.J., Kalra, K.M., Maher, M.M., *et al*. Application of recently developed computer algorithm for automatic classification of unstructured radiology reports: validation study. *Radiology* **324**(2) (2005) 323–329.

[6] Dridi, O. and Ben Ahmed, M. Building an ontology-based framework for semantic information retrieval: application to breast cancer. *Proceedings of 3rd International Conference on Information and Communication Technologies: From Theory to Applications. ICTTA 2008*, 7–11 April, pp. 1–6.

[7] Duh, K. and Kirchhoff, K. Automatic learning of language model structure. *Proceedings of the 20th International Conference on Computational Linguistics (Geneva, Switzerland)*. COLING 2004. ACM Press, New York, 2004, pp. 148–154.

[8] Fickenscher, K.M. The new frontier of data mining. *Health Management Technology* 26(10) (2005) 32–36.

[9] Fox, C. Lexical analysis and stoplists. In *Information Retrieval: Data Structures and Algorithms*, W.B. Frakes and R. Baeza-Yates (eds). Prentice-Hall, Englewood Cliffs, NJ, 1992.

[10] Goldberg, D.E. *Genetic Algorithms in Search, Optimization, and Machine Learning*. Addison-Wesley, New York, 1989.

[11] Wang, J.-Y. and Zhu, Z. Framework of multi-agent information retrieval system based on ontology and its application. *2008 International Conference on Machine Learning and Cybernetics*, 12–15 July, pp. 1615–1620.

[12] Kang, K., Lin, K., Zhou, C. and Guo, F. Domain-specific information retrieval based on improved language model. *Fourth International Conference on Fuzzy Systems and Knowledge Discovery*. FSKD 2007, 24–27 August, pp. 374–378.

[13] Muehlenbein, H. Parallel genetic algorithms, population genetics, and combinatorial optimization. *Proceedings of the Third International Conference on Genetic Algorithms*. Morgan Kaufmann, New York, 1989, pp. 416–421.

[14] Mutalik, P.P., *et al*. Solving combinatorial optimization problems using parallel simulated annealing and parallel genetic algorithms. *Proceedings of the 1992 ACM/SIGAPP Symposium on Applied Computing: Technological Challenges of the 1990s*, pp. 1031–1038.

[15] Nowostawski, M. and Poli, R. Parallel genetic algorithm taxonomy. *Proceedings of the Third International Conference on Knowledge-based Intelligent Information Engineering Systems*, 1999, pp. 88–92.

[16] Patton, M.Q. *Qualitative Evaluation and Research Methods*, 2nd edn. Sage Publications, Inc., Newbury Park, CA, 1990.

[17] Patton, R.M. and Potok, T.E. Adaptive sampling of text documents. *Proceedings of the 13th International Conference on Intelligent and Adaptive Systems and Software Engineering*, July 2004.

[18] Patton, R.M. and Potok, T.E. Characterizing large text corpora using a maximum variation sampling genetic algorithm. *Proceedings of the 8th Annual Conference on Genetic and Evolutionary Computation (GECCO 2006)*. ACM Press, Seattle, WA, 2006, pp. 1877–1878.

[19] Pirkola, A., Keskustalo, H., Leppänen, E., Känsälä, A. and Järvelin, K. Targeted s-gram matching: a novel n-gram matching technique for cross- and monolingual word form variants. *Information Research* 7(2), 2002 [available at http://InformationR.net/ir/7/paper126.html].

[20] Porter, M. An algorithm for suffix stripping. *Program* **14** (1980) 130–137.

[21] Porter Stemming Algorithm. Current Jan. 30, 2004. http://www.tartarus.org/ ~martin/PorterStemmer/

[22] Raghavan, V.V. and Wong, S.K.M. A critical analysis of vector space model for information retrieval. *Journal of the American Society for Information Science* **37**(5) (1986) 279–287.

[23] Reed, J.W., Jiao, Y., Potok, T.E., Klump, B.A., Elmore, M.T. and Hurson, A.R. TF-ICF: A new term weighting scheme for clustering dynamic data streams. *Proceedings of the 5th International Conference on Machine Learning and Applications.* ICMLA 2006, pp. 258–263.

[24] Reed, J.W., Potok, T.E. and Patton, R.M. A multi-agent system for distributed cluster analysis. *Proceedings of the Third International Workshop on Software Engineering for Large-Scale Multi-Agent Systems (SELMAS'04). Workshop in conjunction with the 26th International Conference on Software Engineering, Edinburgh, Scotland.* IEE, pp. 152–155.

[25] Rudolph, G. Convergence analysis of canonical genetic algorithms. *IEEE Transactions on Neural Networks* **5**(1) (1994) 96–101.

[26] Salton, G. *Introduction to Modern Information Retrieval.* McGraw-Hill, New York, 1983.

[27] Sebastiani, F. Machine learning in automated text categorization. *ACM Computing Surveys* **34**(1) (2002) 1–47.

[28] Siddiqui T.J. Integrating notion of agency and semantics in information retrieval: an intelligent multi-agent model. *Proceedings of the 5th International Conference on Intelligent Systems Design and Applications.* ISDA 2005, 8–10 September, pp. 160–165.

[29] Starren, J. and Johnson, S.M. Expressiveness of the Breast Imaging Reporting and Database System (BI-RADS 1997). *Proceedings of the AMIAA Annual Fall Symposium* 1997, pp. 655–659.

[30] Tanese, R. Distributed genetic algorithms for function optimization. Ph.D. thesis, University of Michigan, Department of Computer Science and Engineering, 1989.

[31] Thomas, B.J., Ouellette, H., Halpern, E.F., Rosenthal, D.I. Automated computer-assisted categorization of radiology reports. *AJR* **184** (2005) 687–690.

[32] Thompson, S.K. *Sampling.* John Wiley & Sons, Inc., New York, 1992.

[33] Thompson, S.K. and Seber, G.A.F. *Adaptive Sampling.* John Wiley & Sons, Inc., New York, 1996.

[34] Yan, P., Jiao, Y., Hurson, A.R. and Potok, T.E. Semantic-based information retrieval of biomedical data. *Proceedings of the 21st Annual ACM Symposium on Applied Computing – Semantic-Based Resource Discovery, Retrieval and Composition (RDRC 2006),* 2006, pp. 1700–1704.

5.2

An Interactive Search for Rules in Medical Data using Multiobjective Evolutionary Algorithms

Daniela Zaharie, D. Lungeanu, and Flavia Zamfirache
Faculty of Mathematics and Informatics, West University Of Timisoara, Timiş, Romania

5.2.1 Medical Data Mining

Discovering new and useful knowledge from medical data represents a challenge for any data mining task, due to the heterogeneous nature of data (usually consisting of mixed attributes, i.e. nominal, numerical, and logical, with many erroneous or missing values) and to the requirements to express the knowledge in a medically comprehensible form.

An easily understandable manner of expressing hypotheses extrapolated from data is represented by rules in the form: IF *"some conditions on the values of predicting attributes are true"* THEN *"some conditions on the goal attributes are true"*. If there is only one goal attribute and it specifies a class, then we talk about a classification rule, expressing the possibility that data satisfying the antecedent (IF) condition belongs to the class specified in the consequent (THEN) part. When the goal attributes do not express a class, then we deal

Genetic and Evolutionary Computation: Medical Applications   Edited by Stephen L. Smith and Stefano Cagnoni
© 2011 John Wiley & Sons, Ltd

with prediction rules, expressing hypotheses on the dependence between the antecedent and consequent parts of the rules. Finally, when the potential sets of antecedent and consequent attributes are not previously established, we investigate general association rules expressing co-occurrence of different attribute values. Discovering and selecting rules in data is a search process usually guided by measures quantifying the accuracy, comprehensibility, and interest of the rules. These measures are usually conflicting, i.e. an accurate rule is not necessarily interesting nor easy to read, thus the searching process has to be multicriterial.

Evolutionary algorithms (EAs) proved to be valuable instruments in data mining (Freitas, 2002) and a significant number of papers describe the use of EAs in discovering rules from data (Araujo *et al.*, 2000; Fidelis *et al.*, 2000; Mata *et al.*, 2002; Noda *et al.*, 1999) or in post-processing the set of rules previously extracted by non-evolutionary approaches (Gopalan *et al.*, 2006; Ishibuchi and Yamamoto, 2006). Except for the work of Ishibuchi and Yamamoto, they treat the multi criterial character of the search by aggregating all criteria singly through a pre-specified aggregation function (e.g. a product or weighted sum). Approaches based on multiobjective evolutionary algorithms (MOEAs) have also been proposed (Alatas *et al.*, 2008; Dehuri *et al.*, 2008). However, these approaches do not take into account the user opinion. Since no set of quality criteria can be exhaustive, the user should be involved in the search process.

In this section, we present an approach for evolving rules from data, based on an interactive search. Besides selecting the set of criteria and the sets of potential antecedent and consequent attributes, the user can also intervene in the searching process by marking the uninteresting rules. These marked rules are further used in guiding the search process to avoid generating new rules similar to them.

5.2.2 Measures for Evaluating the Rules Quality

The quality of classification, prediction, or association rules can be quantified using different statistical measures, each capturing specific characteristics of the rules. The rules extracted from data should explain the data, be comprehensible, and contain novel and interesting knowledge, thus the measures of quality are divided into three main classes: (i) accuracy measures (quantifying how well the rules explain the data); (ii) comprehensibility measures (quantifying how easily the rules can be interpreted); and (iii) interestingness measures (quantifying the potential to provide new, previously unknown knowledge).

In order to introduce such measures, let us consider rules having the following general structure: IF $(AT_1, AT_2, \ldots, AT_k)$ $THEN$ $(CT_1, CT_2, \ldots, CT_l)$, where AT_i denotes an antecedent term while CT_j denotes a consequent term. Each term involves one data attribute and is a triplet $\langle a, op, value \rangle$ where a is an attribute, op is an operator (equal, different, in, not in, less than, greater than), and $value$ is a possible value or a set of values for the attribute. A data instance $(a_1, \ldots, a_n)$ satisfies (or is covered by) a given rule if all terms (antecedent and consequent) are true for the attribute values in that instance. If the antecedent terms are all true but there exists at least one consequent term which is false, then the data satisfy only the antecedent part of the rule. Similarly, if all consequent terms are true but at least one antecedent term is false then the data satisfy only the consequent part of the rule. Let us denote by A the event that the antecedent part of the rule is satisfied (disregarding the satisfaction of the consequent) and by C the event that the consequent part of the rule is satisfied (disregarding the satisfaction of the antecedent). Then $P(A, C)$ will be the probability that both the antecedent and consequent parts are satisfied, $P(A)$ will be the probability that the antecedent part is satisfied, and $P(C)$ the probability that the consequent part is satisfied. The negation $\overline{A}$ will denote the event corresponding to the case when the antecedent part is not satisfied. We denote in a similar manner the negation of C and the corresponding probabilities.

5.2.2.1 Accuracy Measures

The most frequent accuracy measures (reflecting the likelihood of rules, given the actual data) are: rule support ($Supp$), confidence ($Conf$), accuracy (Acc), specificity ($Spec$), and sensitivity ($Sens$), as defined by Equation (5.2.1).

$$Supp = P(A, C) \qquad Conf = P(A, C)/P(A) \qquad Acc = P(A, C) + P(\overline{A}, \overline{C})$$
$$Spec = P(\overline{A}, \overline{C})/P(\overline{C}) \qquad Sens = P(A, C)/P(C) \tag{5.2.1}$$

The accuracy, specificity, and sensitivity are mainly used in classification tasks, so they can also be defined using the elements of the confusion matrix (true positive/negative cases, false positive/negative cases). For association and prediction rules, the typically used measures are the support and the confidence. In medical rule mining, these measures should be treated with caution, as high support or even high confidence rules are not necessarily interesting from a medical point of view.

5.2.2.2 Comprehensibility Measures

The readability of a rule is usually related to its length, i.e. the number of related terms, therefore a simple comprehensibility measure is $ch = 1 - (l + k)/n$, where n is the maximum number of terms (in the antecedent and the consequent part) and $l + k$ is the effective number of terms in the rule.

5.2.2.3 Interestingness Measures

There are more than 40 objective measures for evaluating the interestingness of a rule (Ohsaki *et al.*, 2006), so choosing appropriate measures for the actual data characteristics is not a trivial task, as Carvalho *et al.* clearly illustrated in their paper, too (Carvalho *et al.*, 2005). They analyzed the correlation between objective interestingness measures and the real human interest evaluated by experts from each domain, and proposed a ranking of objective quality measures. Not surprisingly, different rankings were obtained for different data sets. For the medical data set they tested, the top three measures were: Phi-coefficient (Φ), odds ratio (OR), and cosine measure (cos), as described in Equation (5.2.2).

$$\Phi = \frac{P(A, C) - P(A)P(C)}{\sqrt{P(A)P(C)(1 - P(A))(1 - P(C))}} \quad OR = \frac{P(A, C)P(\overline{A}, \overline{C})}{P(A, \overline{C})P(\overline{A}, C)}$$

$$cos = \frac{P(A, C)}{\sqrt{P(A)P(C)}} \tag{5.2.2}$$

The cosine measure is similar to the interest (*lift*), as described in Equation (5.2.3). Ohsaki *et al.* (2006) presented a similar study, exclusively oriented towards medical data. A large set of measures (41) were evaluated according to some metacriteria expressing the relationship between the objective measure value given to a rule and the quality label assigned to the same rule by a medical expert. The top three measures obtained by combining the rankings corresponding to two medical sets (one on meningitis and the other on hepatitis) were: *accuracy*, *peculiarity*, and *uncovered negative (UN)*, defined in Equation (5.2.3). As peculiarity can be estimated only in the case of discrete attributes, we included in our analysis only the last one as a measure of interestingness. We can see that UN is the difference between accuracy and support, so maximizing this criterion leads to the maximization of the accuracy and minimization of the support. Thus it favors rules that are not necessarily of high support, but could be interesting. Another measure, used in medical data analysis, especially in epidemiology, is the relative risk (RR), also defined in Equation (5.2.3).

$$lift = \frac{P(A, C)}{P(A)P(C)} \quad UN = P(\overline{A}, \overline{C}) \quad RR = \frac{P(A, C)P(\overline{A})}{P(\overline{A}, C)P(A)} \tag{5.2.3}$$

5.2.3 Evolutionary Approaches in Rules Mining

When designing an EA for rules mining, one has to take into account at least the following aspects: (i) representation of rules; (ii) initialization of the population; (iii) recombination and mutation operators; (iv) rules evaluation and selection. There are two main approaches for rules representation: "Pittsburgh" and "Michigan". In the former approach, each element of the population is a set of rules; thus, it deals well with the interaction between rules and is appropriate when looking for sets of rules defining a classifier. On the other hand, the evolutionary operators are complex and rather difficult to implement (Freitas, 2002). In the latter approach, each element of the population encodes one rule; therefore, all elements have the same structure and the evolutionary operators can be more easily implemented. Although it does not deal well with the rules' interaction, the "Michigan" approach is the most frequently used, especially for prediction and association rules, where the interaction is not as critical as it is in the case of the classification ones.

The evolutionary process starts from an initial population, so choosing appropriate initial rules compatible with the actual data is important. Therefore, most EAs for rules mining start with randomly generated rules, while still trying to involve values present in the data to be mined. The mutation and recombination operators are adapted in order to ensure the rules consistency. The success of an evolutionary rules mining process is highly dependent on the quality measures used to evaluate the population elements, and on the strategy of selecting the elements to be transferred in the next generation. Since the quality of a rule depends on several criteria, there are a large number of variants to compute the fitness value.

The proliferation of the EAs in rules mining is motivated by their capability to deal well with continuous numeric data and be easily adapted for solving the tasks arising at different stages of a rules mining process. For instance, Mata *et al.* proposed an evolutionary algorithm for finding frequent item sets in numeric databases (Mata *et al.*, 2002), its advantage over classical non-evolutionary techniques being that it does not require a previous discretization of continuous data. The characteristic of this algorithm is that the frequent item sets are iteratively discovered by guiding the searching process through a penalization mechanism of the data instances already included in a subset. The fitness function is obtained by aggregating several terms related to: the rule's support; the amplitude of the intervals of values corresponding to attributes; the number of attributes in the item set; and the penalization expressing the ratio of data covered by the current rule which are also included in other item sets.

Other variants, such as those proposed by Gopalan *et al.* (2006), use an EA to post-process a rule set extracted using different techniques. The post-processing aims to discover the accurate and interesting rules in large sets of classification rules. The evolved structures correspond to sets of rules, as in the "Pittsburgh" approach, and the selection process consists of two stages: accurate sets of rules are selected, then from these accurate sets of rules, the most interesting ones are selected. In Ishibuchi and Yamamoto (2006), the EAs are also used for post-processing an existing set of rules, but they deal with fuzzy association rules and use a multiobjective approach. EAs are also used to extract classification rules (Fidelis *et al.*, 2000) and prediction rules (Noda *et al.*, 1999) directly from data (the rules satisfying the comprehensibility and interestingness requirements). For these approaches, the rules' quality is computed by aggregating several accuracy, comprehensibility, and interestingness measures; thus the EA has to deal with a single-objective optimization problem.

Recently, Pareto-based multiobjective algorithms have been used to extract fuzzy association, numeric association, or classification rules (Alatas *et al.*, 2008; Dehuri *et al.*, 2008). However, their common problem is that a single run of the algorithm leads to a set of rules which can be quite large, especially when the number of criteria is large. In order to apply such a technique to medical data, one should pay special attention to the nature of data (usually containing mixed attributes and a significant ratio of missing values) and to the choice of accuracy and interestingness measures.

On the other hand, if the aim is to design tools which can help medical specialists in making decisions, one has to take into account the feedback provided by the specialist and to allow him or her to intervene in the rules discovery process.

5.2.4 An Interactive Multiobjective Evolutionary Algorithm for Rules Mining

We propose to involve the user in the search process, as a predefined aggregation of quality criteria is difficult to find and, moreover, it has been suggested that users can also change their opinion on the rules' quality during the evaluation process itself (Ohsaki *et al.*, 2006). The approach we propose is based on a multiobjective evolutionary algorithm (MOEA) having the general structure described in Algorithm 1.

Algorithm 1 Generic MOEA for rules extraction

1: Initialize a population of m rules
2: Evaluate the population
3: **while** "the maximal number of generations is not reached" **do**
4:     Generate m new rules by crossover
5:     Apply mutation to rules obtained by crossover
6:     Evaluate the new elements
7:     Join the old and the new populations
8:     Select the "best" m rules from the joined population
9: **end while**

5.2.4.1 Rules Encoding

Each element (chromosome) of the population corresponds to a rule and consists of a list of components (genes) corresponding to all attributes in the data set. Each component consists of three fields: ⟨*presence flag, operator, value*⟩. The *presence flag* is a binary value specifying whether the corresponding attribute is involved in the rule (either in its antecedent or the consequent part). In case of binary attributes, this is the only important field. The *operator* allows specification of the condition the attribute should satisfy. We use two possible operators for each type of attribute. In the case of numerical attributes, the possible operators are $\leq$ (coded by 0) and $>$ (coded by 1). For the categorical attributes, the operators are $=$ (coded by 1) and $\neq$ (coded by 0). The *value* field contains the value associated with the attribute.

In all cases, an element is a fixed-length list with mixed values (binary, integer, and real). In the following, the number of attributes is denoted by n. The difference between the antecedent and consequent attributes is made only in the evaluation of an element. In case of the classification rules, the class attribute is not included in the population elements, all attributes being predictive.

5.2.4.2 Reproduction Operators

During each generation, a new population is constructed by crossover and mutation from the current one. By *crossover*, a new rule is constructed starting with two randomly selected rules from the current population. In case of rules containing only terms based on operators as *equal, different, less than,* and *greater than*, the crossover procedure can be described as follows: (i) if the attribute is absent from both parent rules, it will be absent from the generated rule as well; (ii) if the attribute is present in only one parent rule, its operator and value field are transferred to the new rule; (iii) if the attribute is present in both rules and satisfies the same type of condition (the operator field has the same value in both rules), then it is transferred to the new rule (for numerical attributes, the new value is the average of the values corresponding to the parent rules; for nominal attributes, one of the parents' values is just randomly taken); (iv) if the attribute is present in both rules and it satisfies different conditions, then the triplet to be transferred into the new rule is taken from the parent rule having a higher support.

The *mutation* has the role of modifying the rules obtained by crossover. For each attribute, mutation is applied with a given probability (e.g. $p_m = 1/n$) and it can affect one of the fields (i.e. presence flag, operator, or value) and only one at each mutation step. By switching the presence flag, some attributes can be inserted or removed from the rule, thus leading to either a more general or a more specific rule. By changing the operator field, one changes the condition the attribute should satisfy. The mutation of the value field consists of choosing a new value based on a uniform selection from the range of values corresponding to the attribute. If the new element generated by crossover and the mutation is not valid (e.g. it does not contain any antecedent or consequent term), then a repairing rule is applied (e.g. a randomly generated antecedent or consequent term is introduced).

5.2.4.3 Selection and Archiving

After a new population is created by crossover and mutation, a selection step (typical to MOEAs) is applied. Our selection strategy is similar to that used in NSGA-II (Deb and Kumar, 2007), meaning that the elements in the joined population (parents and offsprings) are ranked based on the non-domination relationship. A rule is considered as non-dominated, with respect to rules in a given set, if no other rule in that set is better with respect to all criteria. All the elements that are non-dominated with respect to all elements in the joined population belong to the first non-domination front and have rank 1. Subsequently, the non-dominated elements in the population obtained by ignoring the elements of rank 1 belong to the second non-domination front, and so on. The m elements corresponding to the new generation are selected from the $2m$ ranked elements in their ranks' increasing order. To stimulate the diversity of the resulting Pareto front, a crowding distance is used as a second selection criterion: from two elements having the same rank, the one with a larger crowding distance (suggesting that it belongs to a less crowded region) is selected. The crowding distance can be defined in either the objective or the decision variables space. A particular characteristic of our approach is related to the crowding distance between rules.

We analyzed two types of distances, one expressing the structural differences between rules and the other expressing the difference between the data subsets covered by the rules. In the case of two rules R and R' (encoded by lists of n tuples $t = \langle p, o, v \rangle$, where p denotes the presence flag, o denotes the operator, and v denotes the value), the structural distance is defined by Equation (5.2.4).

$$d_S(R, R') = \frac{\sum_{j=1}^{n} d_j(R, R')}{n} \qquad d_j(R, R') = \begin{cases} 0 & \text{if } p_j = p'_j, \quad o_j = o'_j \\ 1 & \text{if } p_j = p'_j, \quad o_j \neq o'_j \\ 2 & \text{if } p_j \neq p'_j \end{cases} \qquad (5.2.4)$$

where p_j denotes the presence flag and o_j denotes the operator corresponding to the jth attribute. Thus, two rules are considered to be identical from a structural point of view if they contain the same attributes and the terms have identical associated operators.

The distance related to the rule coverage is defined as the cardinal of the subset of data which are either covered by the first rule but are not covered by the second rule, or are covered by the second rule but are not covered by the first rule. Thus, the cover-based distance is given by Equation (5.2.5), where $C(R)$ is the set of data instances which satisfy the rule (are covered by the rule) and Δ denotes the symmetrical difference between two sets.

$$d_C(R, R') = \text{card}(C(R) \Delta C(R')) \qquad (5.2.5)$$

In order to take into account both aspects (i.e. structural and semantic similarity), we used the product of these two distances when computing the crowding value. After a given number of generations, an archive of non-dominated elements is constructed. Not all non-dominated elements from the current population are transferred in the archive, but they are filtered such that both the structural and the cover-based distances between any two elements of the archive are larger than a given threshold (in our analysis we used 0.01).

5.2.4.4 User Guided Evolutionary Search

An interactive search allows the user to interfere with the evolutionary process in order to guide it towards interesting regions of the search space. The idea of permitting the user to interfere with the evolutionary process

---

Algorithm 2 General structure of the interactive variant of the rules' extraction algorithm

---

1: Initialize a population of m rules
2: Evaluate the population
3: **for** stage:=1..maxStages **do**
4:    Execute lines 3–9 in Algorithm 1
5:    Construct the archive of rules
6:    Evaluate the rules in the archive for a test data set
7:    Visualize the rules' archive
8:    Get the user evaluation on the rules in the archive
9:    Process the rules marked by the user as uninteresting:
     • Replace the corresponding elements from the population with newly initialized elements
     • Add the marked rules to the list of prohibited ones
10:   Re-evaluate the current population by taking into account the user evaluation criterion Equation (5.2.6)
11: **end for**

---

has already been explored in the context of multiobjective evolutionary optimization (Deb and Kumar, 2007), by asking the user to provide a so-called "reference point" in the objectives space. However, for a medical specialist it would be rather difficult to provide reference points in the space of quality measures. On the other hand, when (s)he sees a list of rules (s)he can decide which are trivial or potentially interesting. Thus we propose an interactive approach based on lists of rules marked by the user as lacking medical interest.

The overall idea of the interactive process of rules evolving is illustrated in Algorithm 2. In the interactive variant, the search process consists of several stages; at each one, the population is evolved for a given number of generations and the archive of the selected non-dominated rules is provided to the user together with all the objective measures computed for the testing data set (measures not necessarily limited to those used as criteria in the optimization process). In our implementation, we used the following set of measures: support, confidence, accuracy, specificity, sensitivity, comprehensibility, odds ratio, lift, uncovered negative, and relative risk (Ohsaki *et al.*, 2006). Based on these criteria and on a subjective evaluation, the user can decide whether there are uninteresting or incomprehensible rules. Then (s)he can mark these rules and proceed to the next stage of the search. The effect of marking the undesirable rules is twofold: firstly, the population elements corresponding to the marked rules are replaced with randomly initialized elements; secondly, the marked rules are added to a list (L_p) of prohibited rules.

This list can be used to compute a supplementary optimization criterion which expresses the user's evaluation. For a rule R, this criterion is computed as the distance between R and the list L_p of prohibited rules:

$$ue(R) = \min\{d(R, R')| R' \in L_p\} \tag{5.2.6}$$

The distance between rules can be the structural distance (Equation (5.2.4)), the cover-based distance (Equation (5.2.5)), or a combination of them (e.g. their product). By using the user's evaluation as a quality criterion, rules "similar" to those marked as uninteresting have little chance to evolve and survive during the following stages of the evolution. On the other hand, employing the user's evaluation in the search can redirect it towards different regions of the searching space, thus leading to the discovery of new rules.

A possible drawback of using a supplementary optimization criterion is that it usually leads to a larger number of non-dominated elements. In order to avoid this, the list of uninteresting rules can be used to introduce some constraints, i.e. all rules similar to those in the list are considered to be unfeasible. The rules' feasibility is employed when the domination relationship between two rules is checked: if one rule is feasible and the other one is unfeasible, the first rule dominates the second one, disregarding the values of the quality criteria; if both rules are unfeasible or both rules are feasible, the domination is decided based on the values of the optimization criteria.

5.2.5 Experiments in Medical Rules Mining

The interactive variant was implemented such that the user can select the following elements: *rules type* (both classification and prediction rules can be evolved); *lists of attributes* (the user selects the antecedent and consequent attributes); *optimization criteria* (the user can choose an arbitrary subset of measures from those available).

The approach was tested for the medical data sets from the UCI repository (http://mlearn.ics.uci.edu/MLRepository.html) and for a set of obstetrical data collected during one year in a hospital of Obstetrics/Gynaecology. The aim of the experiments was twofold: to validate the ability of the evolutionary approach to discover accurate rules, and to analyze the impact of the user's intervention in the searching process.

Table 5.2.1 Average accuracy of the rule sets generated by different rule-based classifiers and by the evolutionary approach

	ZeroR	CR	DT	JRIP	NNge	OneR	PART	MOEA
Pima	0.65	0.71	0.72	0.75	0.74	0.72	0.74	0.73
Breast	0.65	0.90	0.94	0.94	0.95	0.91	0.94	0.92

In order to validate the ability of the implemented multiobjective evolutionary algorithm to extract reliable rules, we firstly tested it in the case of classification problems. The approach we followed was based on the idea of evolving rules corresponding to one class. Therefore, only the terms in the antecedent part of the rules were evolved. As a class cannot be described by a single rule (usually requiring a set of rules), the MOEA provides a set of reciprocally non-dominated rules, which can be interpreted as describing the class itself (even if it does not cover the entire class). So the rules were evaluated both individually and as a set, using a five fold cross-validation approach. In all tests, the population size was set to 50, the maximal number of generations to 100, and the mutation probability to $1/n$.

The results in Table 5.2.1 were obtained for *Pima Indians Diabetes* and *Wisconsin Breast Cancer* (1991) data sets, based on two optimization criteria: accuracy (*Acc*) and uncovered negative (*UN*). The MOEA outcome is comparable to those obtained by applying other rule-based classifiers implemented in the Weka data mining tool (http://www.cs.waikato.ac.nz/ml/weka/): simple rules classifiers (ZeroR, OneR), conjunctive rules classifier (CR), decision table majority classifier (DT), propositional rule learner based on repeated incremental pruning (JRIP), nearest neighbor-like classifier with non-nested generalized exemplars (NNge), partial decision trees (PART).

5.2.5.1 Impact of User Interaction

Two variants of including the user evaluation in the evolutionary process were implemented and tested: user-criterion and user-constraint. *User-criterion*: For each element, a supplementary optimization criterion is the product between the minimal structural and cover-based distances to all the elements in the list of marked rules. *User-constraint*: An element which is structurally or semantically similar to at least one marked rule is considered unfeasible. The feasibility property is employed when the domination relationship between two rules is checked: one unfeasible rule is always dominated by a feasible one and cannot dominate a feasible one; if both

Table 5.2.2 Breast data set: ranges of the quality measures for rules evolved in three MOEA scenarios. Variant I: no user intervention. Variants II (*user-criterion*) and III (*user-constraint*): at each stage, the user marks the rules having the value of confidence, sensitivity, specificity, or accuracy less than 0.75

Var/ Stage	No. rules	Marked rules	Conf. range	Sens. range	Spec. range	Acc. range	Lift range
I/1	16	–	[0.42, 1]	[0.25, 0.98]	[0.31, 1]	[0.54, 0.93]	[1.24, 2.9]
I/2	6	–	[0.23, 1]	[0.25, 0.55]	[0.88, 1]	[0.81, 0.93]	[2.88, 12.4]
I/3	4	–	[0.97, 1]	[0.28, 0.53]	[0.99, 1]	[0.94, 0.96]	[12.1, 12.4]
II/1	16	10	[0.42, 1]	[0.25, 0.98]	[0.31, 1]	[0.54, 0.93]	[1.24, 2.9]
II/2	36	16	[0.75, 1]	[0.07, 0.98]	[0.97, 1]	[0.92, 0.98]	[10.06, 12.4]
II/3	32	19	[0.21, 1]	[0.21, 0.99]	[0.88, 1]	[0.84, 0.98]	[2.6, 12.4]
III/1	16	10	[0.42, 1]	[0.25, 0.98]	[0.31, 1]	[0.54, 0.93]	[1.24, 2.9]
III/2	15	6	[0.23, 1]	[0.28, 0.98]	[0.83, 1]	[0.81, 0.99]	[2.8, 12.4]
III/3	11	2	[0.75, 1]	[0.27, 0.98]	[0.97, 1]	[0.96, 0.99]	[9.3, 12.4]

rules are either feasible or unfeasible, they are compared according to the optimization criteria.

The impact of these variants on the number of non-dominated rules evolved after three stages is illustrated in Table 5.2.2, for the *Wisconsin Breast Cancer* data set (each stage consisted of 100 generations, the final quality measures lying within large ranges). The user intervention consisted of marking all rules with the quality values smaller than a threshold (e.g. confidence, sensitivity, specificity, or accuracy smaller than 0.75). The results in Table 5.2.2 prove that the user intervention led to rules of higher sensitivity and accuracy. As expected, the Variant II (*user-criterion*) led to a larger number of evolved rules, compared with Variant III (*user-constraint*).

Table 5.2.3 presents examples of the evolved rules for the *Wisconsin Breast Cancer* data set. We note the transformation of rule R2: "Clump thickness $\neq$ 0" AND "Uniformity of cell size $\neq$ 1" discovered in stage 2, into the more specialized rules R3 and R4, which have larger values for confidence and accuracy, but similar (or even smaller) values for sensitivity.

We further used the interactive variants of MOEAs to explore the rules' space when analyzing a set of ordinary obstetrical data (*ObGyn* data set) collected during one year in an Obstetrics/Gynaecology hospital, with the final aim of identifying potential risk patterns for preterm births (i.e. birth before 37 weeks of gestation). The data set contained 2316 instances corresponding to full-term birth cases, and 370 corresponding to preterm birth cases. The

Table 5.2.3 Breast data set: examples of high sensitivity and/or accuracy rules

Var/Stage	Rule	Supp	Conf	Sens	Spec	Acc	Lift
I/1	R1: "Clump thickness ≠ 1"	0.34	0.42	0.98	0.31	0.54	1.24
II/2	R2: "Clump thickness ≠ 0" AND "Uniformity of cell size ≠ 1"	0.07	0.75	0.98	0.97	0.97	9.36
II/3	R3: "Clump thickness ≠ 0" AND "Uniformity of cell size ≠ 1" AND "Uniformity of cell shape ≠ 1"	0.07	0.80	0.98	0.97	0.98	10.06
II/3	R4: "Clump thickness ≠ 0" AND "Uniformity of cell size ≠ 1" AND "Uniformity of cell shape ≠ 1" AND "Bar nuclei ≠ 1"	0.07	0.91	0.91	0.99	0.99	11.36

risk patterns could be expressed either as classification rules (IF "some antecedent conditions are satisfied" THEN class=preterm) or as prediction rules (IF "some antecedent conditions are satisfied" THEN "the gestational age is less than a value").

The list of antecedent attributes was selected based on a preliminary analysis conducted by a medical expert. The selected attributes concerning the mother were: age, body mass index (BMI) prior to the pregnancy, number of previous births (both preterm and on-term), number of abortions, number of miscarriages (either before 16 weeks of pregnancy or within the interval 16–27), total number of pregnancies, number of born children. For all these interval variables, the expert also suggested sub-intervals of values further used in evolving the rules.

Some of the evolved prediction rules selected based on their interestingness measures (lift, uncovered negative, relative risk, and odds ratio) are presented in Table 5.2.4. The medical expert found these rules potentially interesting.

5.2.6 Conclusions

Searching for rules in medical data is a difficult task both because of the data characteristics and because of the lack of a universally valid set of quality measures. When evaluating pieces of knowledge extracted from data, the subjective opinion of the medical specialist is valuable and any rules mining system should take full advantage of such a resource. While there are different ways of making use of the user's feedback in an interactive mining system,

Table 5.2.4 ObGyn data set: examples of rules predicting the gestational age (in weeks) evolved using three optimization criteria: *Spec* × *Sens*, *UN*, and *comprehensibility*. Selection based on the values of interestingness measures (*Lift*, *UN*, *RR*, and *OR*)

Prediction rule	Acc	Lift	UN	RR	OR
IF "no. of abortions $\geq$ 1" THEN "gestation < 36"	0.78	1.38	0.76	1.57	1.46
IF "age < 40" AND "BMI $\notin$ [30, 35]" AND "no. of children $\in$ {1, 2, 3}" AND "no. of previous births before 37 weeks $\in$ {1, 2, 3}" THEN "gestation < 37"	0.76	2.14	0.76	3.29	2.14
IF "age $\in$ [30, 35]" AND "BMI $\leq$ 18.5" AND "no. of children = 0" THEN "gestation < 21"	0.99	1343	0.99	∞	2685
IF "no. of miscarriages $\in$ {2, 3}" THEN "gestation < 28"	0.99	42.63	0.99	66.5	47.8

the alternative solutions we have discussed in this section are based on the idea of penalizing the rules marked by the user as uninteresting and, as a consequence, guiding the search towards different regions of the rules' space. The behavior of such an interactive system depends heavily on the user's feedback, therefore it is difficult to evaluate its effectiveness. However, the tests we conducted illustrated the medically validated potential of an interactive evolutionary approach in exploring the rules' space.

References

Alatas B, Akin E and Karci A 2008 MODENAR: Multi-objective differential evolution algorithm for mining numeric association rules. *Applied Soft Computing* 8(1), 646–656.

Araujo DLA, Lopes HS and Freitas AA 2000 Rule discovery with a parallel genetic algorithm. Proceedings of the Genetic and Evolutionary Computation Conference (GECCO 2000), pp. 89–92.

Carvalho DR, Freitas AA and Ebecken NN 2005 Evaluating the correlation between objective rule interestingness measures and real human interest. Proceedings of the European Conference on Principles and Practice of Knowledge Discovery in Databases (PKDD 2005), LNAI 3721, Springer, pp. 453–461.

Deb K, Agrawal S, Pratab A and Meyarivan T 2000 A fast elitist non-dominated sorting genetic algorithm for multi-objective optimization: NSGA-II. Proceedings of the Conference on Parallel Problem Solving from Nature (PPSN 2000), LNCS 1917, Springer, pp. 849–858.

Deb K and Kumar A 2007 Interactive evolutionary multi-objective optimization and decision-making using reference direction method. Proceedings of the Conference on Genetic and Evolutionary Computation (GECCO 2007), London, pp. 781–788.

Dehuri S, Patnaik S, Ghosh A and Mall R 2008 Application of elitist multi-objective genetic algorithm for classification rule generation. *Applied Soft Computing* **8**(1), 477–487.

Fidelis MV, Lopes HS and Freitas AA 2000 Discovering comprehensible classification rules with a genetic algorithm. Proceedings of the Congress on Evolutionary Computing (CEC 2007), La Jolla, CA, USA, pp. 805–810.

Freitas A 2002 *Data Mining and Knowledge Discovery with Evolutionary Algorithms*. Springer-Verlag, Berlin.

Gopalan J, Alhajj R and Barker K 2006 Discovering accurate and interesting classification rules using genetic algorithms. Proceedings of the International Conference on Data Mining (DMIN 2006), CSREA Press, pp. 389–395.

Ishibuchi H and Yamamoto T 2006 Fuzzy rule selection by multi-objective genetic local search algorithms and rule evaluation measures in data mining. *Fuzzy Sets and Systems* **141**(1), 59–88.

Mata J, Alvarez JL and Riquelme JC 2002 An evolutionary algorithm to discover numeric association rules. Proceedings of the ACM Symposium of Applied Computing (SAC 2002), pp. 590–594.

Noda E, Freitas AA and Lopes HS 1999 Discovering interesting prediction rules with a genetic algorithm. Proceedings of the Congress on Evolutionary Computing (CEC 1999), Washington D.C., USA, pp. 1322–1329.

Ohsaki M, Abe H, Tsumoto S, Yokoi H and Yamaguchi T 2006 Proposal of medical KDD support user interface utilizing rule interestingness measures. Proceedings of ICDMW 2006, IEEE Computer Press, pp. 759–764.

5.3

Genetic Programming for Exploring Medical Data using Visual Spaces

Julio J. Valdés, Alan J. Barton, and Robert Orchard
National Research Council Canada, Ottawa, Ontario, Canada

5.3.1 Introduction

Medical data requires a significant investment of time in order for proper domain knowledge to be acquired and continually updated with the latest available concepts. There are many ways to explore such data in order to increase medical understanding. A visual approach is taken here in order to be able to involve a domain expert within a higher-order mining process. Various aspects of the visual approach are possible. It can be used to aid data understanding *per se* (i.e. Visual Data Mining), or as a facilitator of increasing comprehension of data mining results (i.e. Visual Meta-Mining).

A discussion of the various *Visual Spaces* is presented through two example medical data sets related to breast and colon cancers. In particular, examples of *Data Spaces* and *Semantic Spaces* are given where both genetic programming and classical optimization procedures were used for their construction (for comparison purposes). There are several approaches for non-linear dimensionality reduction, ranging from trying to preserve the inter-data distances to the recovery of meaningful low-dimensional structures hidden in

Genetic and Evolutionary Computation: Medical Applications   Edited by Stephen L. Smith and Stefano Cagnoni
© 2011 the Crown in right of Canada

high-dimensional data (manifolds). Examples are multidimensional scaling and variants (Kruskal, 1964; Sammon, 1969; Young, 1981), principal curves (Hastie, 1984; Hastie and Stuetzle, 1988), locally linear embedding (Roweis and Saul, 1991), Laplacian eigenmaps (Belkin and Niyogi, 2003), generative topographic mapping (Bishop *et al.*, 1998) and stochastic nearest neighbor (Hinton and Roweis, 2003), among others. The presented approach for non-linear dimensionality reduction is not oriented towards a particular kind of genetic programming algorithm due to its generic nature.

5.3.2 Visual Spaces

A *Visual Space* is the tuple $\Upsilon = < \underline{O}, G, B, \Re^m, g_o, l, g_r, b, r >$, where $\underline{O}$ is a relational structure ($\underline{O} = < O, \Gamma^v >$, O is a finite set of objects, and Γ^v is a set of relations); G is a non-empty set of *geometries* representing the different objects and relations; B is a non-empty set of *behaviors* of the objects; $\Re^m \subset \mathbb{R}^m$ is a *metric space* of dimension m (euclidean or not), which will be the actual geometric space associated with the *Visual Space*. The other elements are mappings: $g_o: O \rightarrow G$, $\varphi: O \rightarrow \Re^m$, $g_r: \Gamma^v \rightarrow G$, $b: O \rightarrow B$. Further description follows.

5.3.2.1 Visual Space Realization

There are many ways that a *Visual Space* may be presented for use within data mining and other exploratory endeavors. In particular, Virtual Reality (VR) is suitable for several reasons: (i) VR is *flexible*, it allows the choice of different ways to represent the objects according to human perception differences. (ii) VR allows *immersion*, the user can navigate inside the data and interact with the objects in the world. (iii) VR creates a *living* experience, the user is not a passive observer, but an actor in the world. (iv) VR is *broad and deep*, the user may see the VR world as a whole, or concentrate on details. In addition, it is also very important that its use does not require any special background knowledge. A VR technique for visual data mining on heterogeneous, imprecise, and incomplete information systems was introduced by Valdés (2002, 2003).

5.3.2.2 Visual Space Taxonomy

The property(ies) of the objects within a constructed *Visual Space* may satisfy one of the following paradigms, which are with respect to a decision attribute (Valdés and Barton, 2006). Object locations in the *Visual Space* should preserve:

- A structural property of the data (*Unsupervised Paradigm*), where locations are dependent only on the set of descriptor attributes. Data structure is one of the most important elements to consider when the location and adjacency relationships between the objects should give an indication of the *similarity relationships* (Borg and Lingoes, 1987; Chandon and Pinson, 1981) between the objects as given by the set of original attributes (Valdés, 2003).
- A property of the class information (*Supervised Paradigm*), where objects are maximally discriminated with respect to class information. The structure of the space is distorted in order to maximize class discrimination (Valdés and Barton, 2005).
- A compromise between structure and class preservation (*Mixed Paradigm*), where very often these two goals are conflicting (Valdés and Barton, 2006).

5.3.2.3 Visual Space Geometries

Many geometries may be placed within a *Visual Space*. Examples include: (i) *Object Geometries*, where aspects of particular objects are represented. For example, an object could be represented by a sphere, cube, cone, etc. (ii) *Group Geometries*, where aspects of sets of particular objects are represented. For example, the number of objects within the set could be used to appropriately scale the geometry or groups of objects may be represented by a single geometry, such as a convex hull. (iii) *Space Geometries*, where aspects of the complete *Visual Space* are modified in order to support a particular interpretation. For example, a *Visual Space* may be constructed within an *Unsupervised Paradigm*, but then *Supervised* information may be applied to the *Visual Space* through the use of a specific colour applied to each object depending on its class membership. If supervised information is not available, then it may be constructed, through, for example, the use of clustering or statistical means.

5.3.2.4 Visual Space Interpretation Taxonomy

In order to interpret the characteristics of a particular space, it is important to understand how to interpret the underlying meaning of the objects. That is, the origin of the objects is required information for a particular interpretation. Examples of such spaces are as follows.

- *Data Spaces*: The data from a particular domain is used to construct a space. *Data Spaces* are an interpretation of the data themselves and are not related to any particular algorithm (e.g. machine learning algorithm, data mining algorithm, etc.). See Valdés *et al.* (2007a) for an example of a *Data Space*.

- *Parameter Spaces*: The parameters controlling an investigated algorithm are used to construct a space. *Parameter Spaces* are an interpretation of the properties of the *input* to an investigated algorithm. For example, an Evolutionary Computation-based algorithm such as Genetic Programming may have parameters such as maximum number of generations, set of basis functions, etc. See Barton (2009) for an example of a *Parameter Space*.
- *Semantic Spaces*: The results from an investigated algorithm are used to construct a space for meta-mining purposes. *Semantic Spaces* are an interpretation of the properties of the *results* (i.e. the output) of an investigated algorithm. Examples include the following.

  - *Variable Spaces*, where the original objects are related to the *variables* within the GP constructed equation(s).
  - *Function Spaces* are similar to *Variable Spaces*, but *function* information is extracted from the GP equation(s) instead of variable information.
  - *Rule Spaces*, where the original objects are rules. For example, meta-mining (Abraham and Roddick, 1999; De Preter *et al.*, 2009) produces meta-rules, which are patterns within the rules found by a data-mining algorithm. Here a *Rule Space* (Valdés, 2003) is a *Visual Space* rather than, for example, a set of meta-rules.
  - *Objective Spaces*, where the original objects are composed of a set of objective function values obtained from, for example, a Multi-Objective Optimization procedure (Valdés *et al.*, 2007b).

More generally, different information sources are associated with the attributes, relations, and functions. They are described by source sets (Ψ_i), constructed according to the nature of the information represent. Source sets also account for incomplete information. A heterogeneous domain (Valdés, 2003) is a Cartesian product of a collection of source sets: $\hat{\mathcal{H}}^n = \Psi_1 \times \cdots \times \Psi_n$, where $n > 0$ is the number of information sources to consider. For example, in a domain where objects are described by continuous crisp quantities, discrete features, fuzzy features, time series, images, and graphs, they can be represented as Cartesian products of subsets of real numbers ($\hat{R}$), nominal ($\hat{N}$), or ordinal sets ($\hat{O}$), fuzzy sets ($\hat{F}$), sets of images ($\hat{I}$), sets of time series ($\hat{S}$), and sets of graphs ($\hat{G}$), respectively (all extended to allow for missing values). The heterogeneous domain is $\hat{\mathcal{H}}^n = \hat{N}^{n_N} \times \hat{O}^{n_O} \times \hat{R}^{n_R} \times \hat{F}^{n_F} \times \hat{I}^{n_I} \times \hat{S}^{n_S} \times \hat{G}^{n_G}$, where n_N is the number of nominal sets, n_O of ordinal sets, n_R of real-valued sets, n_F of fuzzy sets , n_I of image-valued sets, n_S of time-series sets, and n_G of graph-valued sets, respectively ($n = n_N + n_O + n_R + n_F + n_I + n_S + n_G$).

5.3.2.5 Visual Space Characteristics Examination

The examination of a particular paradigm of *Visual Space* uses a definite interpretation. For example, a *Function Space* constructed within an *Unsupervised Paradigm* yields visual information about the relationships between the set of functions used to describe the objects. Therefore, the inspector of such a space directly examines the characteristics, which may involve incorporating domain information or other types of knowledge. However, in addition, it may be of interest to use an algorithm in order to formalize the objective construction of such *Visual Space* characteristics.

Rough Sets for General Visual Space Characterization

Rough Set Theory (Pawlak, 1991) is based on the assumption that in order to define a set, some knowledge about the elements of the set is needed. Some elements may be indiscernible from the point of view of the available information, and knowledge is understood to be the ability to characterize all classes of a classification of the set elements. An information system is a pair $\mathbf{A} = (U, A)$, where U is a non-empty finite set (the universe) and A is a non-empty finite set of attributes. A decision table is any information system of the form $\mathbf{A} = (U, A \cup \{d\})$, where $d \in A$ is the decision attribute and the elements of A are the condition attributes.

An information system $\mathcal{A}$ defines a matrix M_A called a discernibility matrix. Each entry consists of the set of attributes that can be used to discern between objects. M_A defines a binary relation $R_A \subseteq U^2$ called an *indiscernibility relation* with respect to A, and expresses which pairs of objects cannot be discerned. If an attribute subset $B \subseteq A$ preserves the indiscernibility relation R_A, then the attributes $A \backslash B$ are said to be *dispensable*. An information system may have many such attribute subsets B. All such subsets that are minimal (i.e. that do not contain any dispensable attributes) are called reducts. The set of all reducts of an information system $\mathcal{A}$ is denoted $RED(\mathcal{A})$. In particular, minimum reducts (those with a small number of attributes) are extremely important, as decision rules can be constructed from them (Bazan *et al.*, 1994). The problem of reduct computation is NP-hard, and several heuristics have been proposed (Wróblewski, 2001).

Reducts and reducts-derived decision rules were used to produce additional characterization of the *Semantic Spaces* derived from meta-mining GP results.

5.3.2.6 *Visual Space Mapping Taxonomy*

One of the steps in the construction of a *Visual Space* for data representation is the transformation of the original set of attributes describing the objects under study, often defining a heterogeneous high-dimensional space, into another space of small dimension (typically two to three) with an intuitive metric (e.g. Euclidean). The operation usually involves a non-linear transformation, implying some information loss. From the point of view of their mathematical nature, the mappings from the data to the *Visual Space* can be as follows.

- *Implicit*: where the images of the transformed objects are computed directly with no associated mapping function representation.
- *Explicit*: where the function performing the mapping is found by the procedure and the images of the objects are obtained by applying the function. Two functional representation sub-types are:
  - *analytical functions*, e.g. $3 + v_1 \cdot (4.1 + v_3)$.
  - *general function approximators*, e.g. neural networks (Barton, 2009), fuzzy systems, or others.

Let $\mathcal{H}^n$ be the space defined by the original n-attributes (not necessarily numeric) and $\mathbb{R}^m$ an m-dimensional space given by a Cartesian product of the reals s.t. $m < n$. A mapping $\varphi:\mathcal{H}^n \rightarrow \mathbb{R}^m$ creates images of the objects in $\mathcal{H}^n$ in $\mathbb{R}^m$. The mappings of interest are those which maximize some metric or non-metric structure preservation criteria, as has been done for decades in multidimensional scaling (Borg and Lingoes, 1987; Kruskal, 1964), or to minimize some error measure of information loss (Sammon, 1969). If δ_{ij} is a dissimilarity measure between any two objects i,j ($i,j \in \mathcal{H}^n$), and $\zeta_{i^v j^v}$ is another dissimilarity measure defined on objects $i^v, j^v \in \mathbb{R}^m$ ($i^v = \varphi(i), j^v = \varphi(j)$), examples of error measures frequently used are:

$$S\ stress = \sqrt{\frac{\sum_{i<j} (\delta_{ij}^2 - \zeta_{ij}^2)^2}{\sum_{i<j} \delta_{ij}^4}} \qquad (5.3.1)$$

$$Sammon\ error = \frac{1}{\sum_{i<j} \delta_{ij}} \frac{\sum_{i<j} (\delta_{ij} - \zeta_{ij})^2}{\delta_{ij}} \qquad (5.3.2)$$

$$Quadratic\ loss = \sum_{i<j} (\delta_{ij} - \zeta_{ij})^2 \qquad (5.3.3)$$

5.3.2.7 Visual Space Mapping Computation

Explicit mappings can be constructed in the form of analytical functions (e.g. via genetic programming), or using general function approximators like neural networks (Barton, 2009) or fuzzy systems. An explicit mapping (e.g. φ) is useful for both practical and theoretical reasons. On the one hand, in dynamic data sets (e.g. systems being monitored or incremental data bases) an explicit transform φ will increase the update rate of the VR information system. On the other hand, it can give semantics to the attributes of the VR space, thus acting as a general dimensionality reducer.

Classical algorithms have been used for directly optimizing these measures, like steepest descent, conjugate gradient Fletcher–Reeves, Powell, Levenberg–Marquardt, and others. The number of different similarity, dissimilarity, and distance functions definable for the different kinds of source sets is immense. Moreover, similarities and distances can be transformed into dissimilarities according to a wide variety of schemes, thus providing a rich framework.

For this study, in addition to genetic programming, the well-known deterministic optimization technique of the Fletcher–Reeves–Polak–Ribiere method (FRPR) was used (Press *et al.*, 2000). It assumes that the function f is roughly approximated as a quadratic form in the neighborhood of an N-dimensional point **P** and uses the information given by the partial derivatives of the original function f. This is the conjugate gradient family of minimization methods and requires an initial approximation to the solution (typically random), which is then refined in a sequence of iterative steps. The convergence of these methods is relatively fast, but they suffer from the entrapment effect, by means of which the obtained solutions are locally optimal.

Visual Space Mapping Computation using Genetic Programming

Genetic programming techniques aim at evolving computer programs, which ultimately are functions. Genetic programming is an extension of the genetic algorithm introduced Koza (1989) and further elaborated in Koza (1992, 1994) and Koza *et al.* (1999). Genetic programming combines the expressive high-level symbolic representations of computer programs with the near-optimal search efficiency of the genetic algorithm. For a given problem, this process often results in a computer program which solves it exactly, or if not, at least provides a fairly good approximation. Those programs which represent functions are of particular interest and can be modeled as $y = F(x_1, \cdots, x_n)$, where $(x_1, \cdots, x_n)$ is the set of independent or predictor variables, and y the dependent or predicted variable, so that $x_1, \cdots, x_n, y \in \mathbb{R}$, where $\mathbb{R}$ are the reals. The function F is built by assembling functional subtrees using a set of predefined primitive functions (the function set), defined beforehand. In general terms, the model describing the program is given by $y = F(\vec{x})$, where $y \in \mathbb{R}$ and $\vec{x} \in \mathbb{R}^n$. Most implementations of genetic programming for modeling fall within this paradigm.

However, there are problems involving vector functions that require an extension. In these cases the model associated with the evolved programs must be $\vec{y} = F(\vec{x})$. Note that these are *not* multi-objective problems, but problems where the fitness function depends on vector variables. The mapping problem between vectors of two spaces of different dimension (n and m) is one of that kind. In this case a transformation like $\varphi: \mathbb{R}^n \rightarrow \mathbb{R}^m$ mapping vectors $\vec{x} \in \mathbb{R}^n$ to *vectors* $\vec{y} \in \mathbb{R}^m$ would allow a reformulation of Equation (5.3.2) (and others) as in Equation (5.3.4):

$$\text{Sammon error} = \frac{1}{\sum_{i<j} \delta_{ij}} \sum_{i<j} \frac{(\delta_{ij} - d(\vec{y}_i, \vec{y}_j))^2}{\delta_{ij}} \tag{5.3.4}$$

where $\vec{y}_i = \varphi(\vec{x}_i)$, $\vec{y}_j = \varphi(\vec{x}_j)$. The implication from the point of view of genetic programming is that instead of evolving expression trees, where there is a one-to-one correspondence between an expression tree and a fitness function evaluation (the classical case), the evolution has to consider populations of *forests* such that the evaluation of the fitness function depends on the set of trees within a forest. In these cases, the cardinality of any forest within the population is equal to the dimension of the target space m.

The variant of GP used is called Gene Expression Programming (GEP) (Ferreira, 2006). In the GEP algorithm, individuals are encoded as simple strings of fixed length, referred to as chromosomes. Each chromosome can be composed of one or more genes which hold individual mathematical expressions that are linked together to form a larger expression. To facilitate experimentation with the GEP algorithm, an extension to an existing Java-based Evolution Computing Research System called ECJ (ECJ, 2007) from George Mason University was made that implements four basic types of problem: (i) function finding, (ii) classification, (iii) time series, and (iv) logical. It supports a large set of fitness functions and allows expressions to be created using many types of arithmetic functions. Batch files may be constructed (outside of ECJ/GEP) for execution upon distributed systems (e.g. Condor: http://www.cs.wisc.edu/condor/ University of Wisconsin-Madison) to allow for large-scale experimentation.

Importance Functions for Variable Space and Function Space Characterization

Each individual GP experiment produces a model in the form of a vector with a number of functions equal to the dimension of the target space, and a fitness measure (typically error related). When a collection of models are considered (like those resulting from a distributed computing experiment), important information can be obtained when their composition is considered from a statistical perspective. A natural approach is to use the empirical distribution functions of predictor variables within the set of equations defining a model (the same applies (*mutatis mutandis*) to the functions from the function set and to a hybrid vector composed of both variables and functions).

Let M be the number of GP models obtained from a distributed computing experiment involving N_V independent (predictor) variables and a function set of cardinality N_F. Let $N_{i,j}$ be the actual number of occurrences of the jth variable in the ith GP model and let f_i be the fitness value corresponding to the ith model (in a [0,1] range). An importance measure for the occurrence of the given variable within a model can be expressed as the weighted term $I_{ij} = f_i N_{i,j}$ and by $\sum_{i=1}^{M}(f_i N_{i,j})$ when a set of M models is considered. The terms I_{ij} define a matrix called the importance matrix (high frequency of occurrence in a high fitness model implies high importance), composed of vectors $\overleftarrow{I_i}$ of dimension N_V. A relative measure can be obtained by normalizing the column (variable) marginals by the overall number of total occurrences of the whole set of predictor variables, assuming that all models would have been a perfect fit (i.e. $f_i = 1$ for all $i \in [1,M]$):

$$I_{ij} = f_i N_{i,j} \tag{5.3.5}$$

$$\mathcal{I}_j = \frac{\sum_{i=1}^{M} I_{ij}}{\left(\sum_{i=1}^{M} \sum_{j=1}^{N_V} N_{i,j}\right)} \tag{5.3.6}$$

5.3.3 Experimental Settings

Two data sets (Breast Cancer and Colon Cancer) were used along with both a classical and a genetic programming technique for the construction of visual spaces. The experimental settings associated with each such experimental condition are detailed subsequently.

The University of California, Irvine (UCI) maintains an international machine learning (ML) database repository (http://www.ics.uci.edu/~mlearn/), containing an archive of over 100 databases used specifically for evaluating machine learning algorithms. As such, a breast cancer database created by Dr William H. Wolberg of the University of Wisconsin Hospitals, Madison, Wisconsin, was obtained from the UCI ML Repository (Mangasarian and Wolberg, 1990; Wolberg and Mangasarian, 1990). The database contains 699 objects, nine integer-valued attributes, and is composed of two classes (241 malignant and 458 benign). In addition, 16 objects contain missing values, which, for the purposes of our study, were removed from consideration (leaving 683 objects).

A previous study (Whitney *et al.*, 2006) analyzed a human colon cancer cell line (called EcR-RKO/KLF4) that was treated for up to 24 hours with ponasterone A to induce expression of a full-length, transgenic Krüppel-like factor 4 (KLF4). KLF4 is an epithelially enriched, zinc finger transcription factor. The study's results provide insight into the biochemical function of KLF4. This cell cycle (Wikipedia, 2007) is the series of events in a eukaryotic cell between one cell division and the next. Thus, it is the process by which a single-cell fertilized egg develops into a mature organism and the process by which hair, skin, blood cells, and some internal organs are renewed. It consists of four distinct phases: G1 phase, S phase, G2 phase (collectively known as interphase), and M phase. M phase is itself composed of two tightly coupled processes: meiosis, in which the cell's chromosomes are divided between the two daughter cells, and cytokinesis, in which the cell's cytoplasm physically divides. Cells that have temporarily or reversibly stopped dividing are said to have entered a state of quiescence called G0 phase, while cells that have permanently stopped dividing due to age or accumulated DNA damage are said to be senescent. The molecular events that control the cell cycle are ordered and directional; that is, each process occurs in a sequential fashion and it is impossible to "reverse" the cycle. Data was obtained from   http://www.ncbi.nlm.nih.gov/geo/gds/gds_browse.cgi?gds=1942 and consists of 44,566 objects, eight time points (0h, 1h, 2h, 4h, 6h, 8h, 12h, 24h) (only the last seven used), and two classes (22,283 control and 22,283 induced). After application of the classical leader algorithm (Hartigan, 1975), the class distribution was 224 control and 335 induced for a total of 559 objects.

5.3.3.1 Implicit Classical Algorithm Settings

The FRPR method (a slight modification of this method, believed to improve the original) is the one used here as a standard for comparison. When Equation (5.3.4) is solved for the $\vec{y}_k \in \mathbb{R}^m$, $k \in [1, N]$, where N is the number of objects, with this method (or a similar one), the obtained mapping is implicit. In order to reduce the chance of entrapment in local minima, 100 runs of the algorithm were made, that were started with the corresponding number of random initial approximations (in this case random 3D matrices). This procedure was applied to each investigated data set. The solutions were used to construct the corresponding virtual reality spaces with which those obtained by vectorial genetic programming are compared.

5.3.3.2 Explicit GEP Algorithm Settings

The major experimental settings that were used for the ECJ/GEP experiments are listed in Table 5.3.1. It can be seen that the mathematical expressions

Table 5.3.1 Experimental settings used for the investigation of the Breast and Colon Cancer data

GEP Parameter	Experimental Values
Number of Generations	100, 500, and 1000
Population Size	100, 200, and 500
Number of Chromosomes/Individual	3
Number of Genes/Chromosome	5
Gene Head Size	5
Linking Function	Addition
Number of Constants/Gene	4
Bounded Range of Constants	[1, 10]
Inversion Rate	0.1
Mutation Rate	0.044
is-transposition Rate	0.1
ris-transposition Rate	0.1
One-point Recombination Rate	0.3
Two-point Recombination Rate	0.3
Gene Recombination Rate	0.1
Gene Transposition Rate	0.1
rnc-mutation Rate	0.01
dc-mutation Rate	0.044
dc-inversion Rate	0.1
Breast Cancer Data	
5 Functions(weight)	add(2), sub(2), mult(2), div(1), pow(1)
Breast and Colon Cancer Data	
8 Functions(weight)	add(2), sub(2), mult(2), div(1), pow(1), exp(1), sin(1), cos(1)

Table 5.3.2   Example of a mathematical
expression and its respective encoding

Mathematical expression:	b*(c − d)
Karva encoding of expression:	*b − cd/bdaac

composed for each of the chromosomes in an individual will be made from five gene expressions, each linked by the addition operator. The gene expressions can be formed using constants, the independent variables associated with each problem, and any of the functions listed. The weightings for the functions allow us to give a preference to some operators over others. In this case we decided to favor addition, subtraction, and multiplication. The number of *characters* (constants, functions, variables) in each gene is determined by the gene's head size and the maximum arity of any function being used, and consequently is calculated according to Equation (5.3.7):

$$headSize + headSize * (maximumArity - 1) + 1 \qquad (5.3.7)$$

In this study a head size of 5 and a maximum arity of 2 were used, leading to genes that have 11 characters that encode expressions in a special notation called Karva (Ferreira, 2006). Each gene also stores four constants randomly chosen between 1 and 10 that can be used in the formation of Karva expressions. A Karva expression with 11 characters and the expression it encodes is shown in Table 5.3.2; depending on the expression being encoded, not all characters are part of the expression.

The great advantage of the encoding scheme is that it is compact and any combination of the allowed characters will always represent a valid mathematical expression.

The parameters related to evolution determine how the genes (Karva expressions) are transformed from generation to generation. It is beyond our scope here to discuss this in detail, but as an example, the mutation rate determines the number of characters that will be mutated (replaced by another character) in the population. If we have 100 individuals in the population and each individual has one chromosome with a single gene of size 11 characters, then we would randomly select $100 * 11 * 0.044$ or 48 characters to mutate. Note that it may be that the character is in a region of the gene that is not part of the expression encoded by the gene, and the effect will then be no change to the expression. The parameters rnc-mutation rate, dc-mutation rate, and dc-inversion rate control how the four constants associated with each gene evolve from generation to generation, while the others are related to evolution of the genes.

For each of the three experiments a set of 75 runs was performed with random seeds and varying values for the number of generations and population size.

5.3.4 Medical Examples

Two medical data sets are used in order to demonstrate examples of *Visual Spaces*. VR is used, which has a dynamic and interactive nature that is not representable on static media. However, snapshots of selected perspectives are possible and have been taken in order to report *Data Spaces* and *Semantic Spaces* within this work.

5.3.4.1 Data Space Examples

A previous examination of Colon Cancer and Breast Cancer data was performed in which a large number of examples of *Data Spaces* were reported (Valdés *et al.*, 2007a). The best Sammon errors (see Equation (5.3.4)) for each of the five groups within the experimental design are shown in Table 5.3.3. For the Breast Cancer data, the FRPR method outperformed GP/GEP by a factor of 2.6, while for Colon Cancer data, GP/GEP outperformed FRPR by a factor of 10. Further investigation is required in order to more thoroughly understand the properties of each method. As a first thrust in that direction, the error histograms associated with the two groups of colon cancer

Table 5.3.3   Best Sammon error result found by each method for computation of a *Data Space*.

Medical Data Set	Method		Best Sammon Error
1) Breast Cancer	GP/GEP (5 functions)	225 Experiments	0.026616
2)	GP/GEP (8 functions)	225 Experiments	0.027823
	FRPR	100 Trials	0.009945
3) Colon Cancer	GP/GEP (8 functions)	225 Experiments	0.002874
	FRPR	100 Trials	0.024015

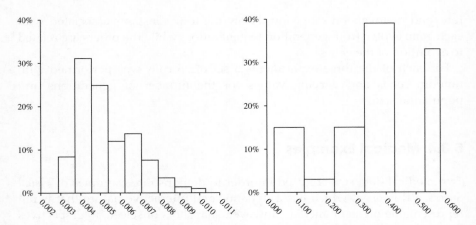

Figure 5.3.1 Sammon error histograms for colon cancer. Left: GP/GEP (8 functions). Right: FRPR

experiments were selected (see Figure 5.3.1). It can clearly be seen that the GP/GEP method has a positively skewed distribution, while the FRPR tends towards that of negative skewness with higher Sammon error values; indicating the advantage of GP/GEP on this data.

Due to space limitations, the equations associated with two of the three (see Valdés *et al.*, 2007a for the other) best Sammon error GP/GEP experiments in Table 5.3.3 are as follows.

- The best Breast Cancer GP/GEP experiment using five functions yielded the vector function mapping (φ) of Equation (5.3.8):

$$
\begin{aligned}
\varphi_X &= v2 + v6 + v3 + ((v3/v2) - v3) \\
          &\quad + ((v6 - (v9 + v6))/v5) \\
\varphi_Y &= (v7/((v1 + v4) + k_{y1})) + ((v7 * v3)/(v3 + v3)) \\
          &\quad + (v4 - v7) + (v7/((v2 + v3) * pow(v6, v4))) + v7 \\
\varphi_Z &= v5 + (v6/(k_{z1}/v9)) + ((v6 + v6)/(v5 + v9)) \\
          &\quad + (v8 - v5) + (v1 - v6)
\end{aligned} \tag{5.3.8}
$$

where $k_{y1} = 3.73$, and $k_{z1} = 9.95$.

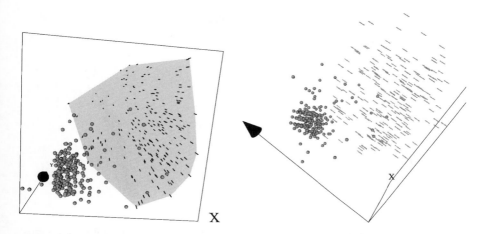

Figure 5.3.2 *Data Space* for Breast Cancer. Gray spheres: Benign class. Black rods: Malignant class. A semitransparent gray membrane wraps the Malignant class. Left: GP result with five functions in the function set (mapping error = 0.026616) (see Equation (5.3.8)). Right: FRPR result (mapping error = 0.009945)

- The best Colon Cancer GP/GEP experiment using eight functions yielded the vector function mapping (φ) of Equation (5.3.9):

$$
\begin{aligned}
\varphi_X &= (v6 - ((v2 - v5)/k_{x1})) + v4 + exp(cos(v1)) \\
&\quad + cos(v3) \\
\varphi_Y &= ((v5 + v3)/(k_{y1} - v7)) + v1 + cos(v1) + (v2/k_{y2}) \\
&\quad + (v2/v1) \\
\varphi_Z &= cos(v5) + k_{z1} + sin((cos(v7) * (v7 * k_{z2}))) \\
&\quad + ((v3 - v6)/k_{z3}) + v3
\end{aligned}
\tag{5.3.9}
$$

where $k_{x1} = 3.65$, $k_{y1} = 3.21$, $k_{y2} = 1.92$, $k_{z1} = 1.22$, $k_{z2} = 6.18$, $k_{z3} = 6.86$.

Each of these equations was then used to construct a *Data Space* along with the FRPR method for such *Data Space* construction. The Breast Cancer examples are shown in Figure 5.3.2 for Equation (5.3.8) and the Colon Cancer examples are shown in Figure 5.3.3 for Equation (5.3.9).

At the global scale for the Breast Cancer *Data Spaces* shown in Figure 5.3.2, it can be seen that the higher Sammon error result for GP contains the overall data structure of that for the lower Sammon error result for FRPR. That is, samples coming from the Malignant class are more diverse and to the right

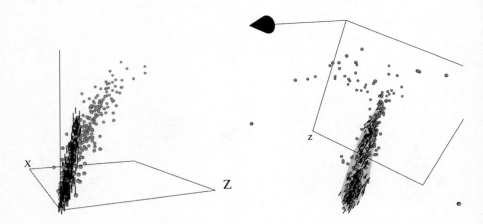

Figure 5.3.3 *Data Space* for Colon Cancer. Gray spheres: Control class. Black rods: Induced class. Left: GP result with eight functions in the function set (mapping error = 0.002874) (see Equation (5.3.9)). Right: FRPR result (mapping error = 0.024015)

of the other class. For some applications, this may be sufficient because the advantage of the GP result is the explicit function that is obtained for such a mapping vs. the implicit (i.e. unknown) function obtained by FRPR. The local object placement differences between the two methods is the reason for the difference in the overall Sammon error value.

For the Colon Cancer *Data Spaces* shown in Figure 5.3.3 the roles of the two methods are reversed. That is, GP found a lower Sammon error solution than FRPR. It is difficult to see in the figure, but the FRPR result is curved from front to back and then splits like a tree at the top, while the GP result is not curved backwards. Overall, for both spaces, the Induced class lies mainly to the left of the Control class, indicating agreement of the two methods.

5.3.4.2 Semantic Space Examples

In order to illustrate the use of semantic spaces for visual meta-mining, examples are presented for both Breast and Colon Cancer cases. As explained in Section 5.3.2.7, importance matrices were computed for the variables and functions composing the 225 GP models obtained for each case. For Breast Cancer there were nine predictor variables and $\{5, 8\}$ functions in the function set, whereas for Colon Cancer there were seven predictor variables and eight functions in the function set. Importance matrices (see Equation (5.3.5)) were computed and their associated unsupervised spaces were obtained. The models were classified according to their mapping error values into three equal probability classes, using the 0.33 and 0.66 quartiles of the error distribution as class boundaries. The importance measures for the variables of both examples in the sense of Equation (5.3.6) were also computed.

Visual spaces of dimension 3 were generated using a dissimilarity given by $1/(1 + s)$ where s is Gower's similarity coefficient (Gower, 1973) and Euclidean distance in the visual space. The space transformation was computed with the FRPR. Then, each model was represented with a geometry and color indicating its error class (low, medium or high, inverse to model quality).

Since the purpose of this section is to illustrate the use of visual spaces in meta-mining collections of GP models obtained from the same data, not all possible semantic spaces for the two medical examples will be presented.

Variable Importance: Colon and Breast Cancer

The importance measures for the variables (Equation (5.3.6)) of the Colon and Breast Cancer examples are shown in Figure 5.3.4. In the Colon Cancer case all variables have a substantial contribution from the point of view of model composition, with a higher importance for variables 3 and 7. In the Breast Cancer case, there are more acute differences when models with eight functions in the function set are considered, with variables $\{6, 8, 9\}$ having higher relative importance and $\{1, 3\}$ with little incidence. The relative differences are not so acute with five functions in the function set, but variables $\{6, 9\}$ outstand. Although preliminary, this approach is a complement to other techniques for outline variable importance from a feature selection

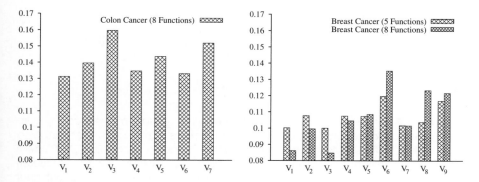

Figure 5.3.4 Variable importance for each data set. Left: Colon Cancer. Right: Breast Cancer.

perspective, as well as a promising one, since it focuses on the actual use of the different variables by the GP model search mechanism (also considering model quality).

Variable Spaces: Breast Cancer

The space corresponding to the nine-dimensional importance matrix associated with the set of GP models for the Breast Cancer data with a cardinality 5 function set is shown in Figure 5.3.5. Overall, there is no differentiation between the low, medium, and high error classes (rods, cones, and spheres, respectively). From another perspective, rough set reduct computation (see Section 5.3.2.5) (using Johnson's algorithm) produced a single reduct composed of all nine variables from which 214 decision rules were generated. This indicates that no clear relationships can be established, as the visual inspection of Figure 5.3.5 shows. However, there is a region of higher model density close to the origin for all class models (particularly for low error ones).

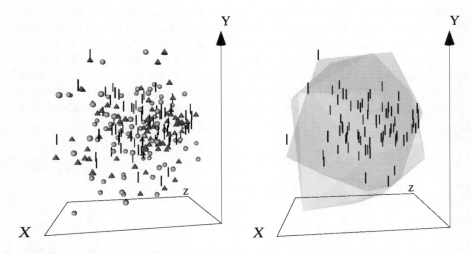

Figure 5.3.5 Breast Cancer: *Semantic Space* for the variable importance measure corresponding to the 225 GP models with five functions in the function set (mapping error = 0.052974). Left: Black rods = models with mapping error < 0.032255 (lower error class), gray cones = models with mapping error within [0.032255, 0.038053] (medium error class), light spheres = models with mapping error ≥ 0.038053 (higher error class). Right: Low error class (rods) with semitransparent membranes wrapping the medium and higher error classes for comparison

It is clearly visible in Figure 5.3.5(right) when the wrapper surfaces for the medium and higher error classes are used.

Function Spaces: Colon and Breast Cancer

The space corresponding to the eight-dimensional function importance matrix associated with the set of GP models for the Colon Cancer data is shown in Figure 5.3.6. It shows a high density core of models (predominantly low and medium error models), surrounded by a lower density shell of outlying elements (predominantly higher error models). The wrapped representation of the medium and higher error classes (Figure 5.3.6(right)) presents a simplified view of the same behavior.

In the Breast Cancer case, the semantic spaces for the GP models with both five and eight functions exhibit similar behavior (Figures 5.3.7 and 5.3.8). The concentration of low and medium error models increases along the X axis, which the right-hand side of both figures shows clearly when the wrappers for

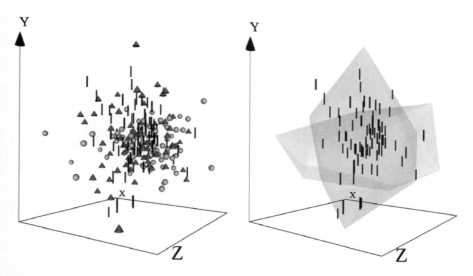

Figure 5.3.6 Colon Cancer: *Semantic Space* for the function importance measure corresponding to the 225 GP models with eight functions in the function set (mapping error = 0.051392). Left: Black rods = models with mapping error < 0.004483 (lower error class), gray cones = models with mapping error within [0.004483, 0.005711) (medium error class), light spheres = models with mapping error ≥ 0.005711 (higher error class). Right: Low error class (rods) with semitransparent membranes wrapping the medium and higher error classes for comparison

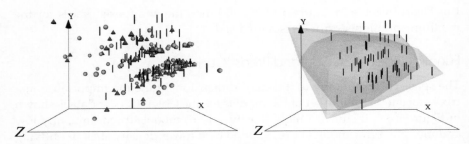

Figure 5.3.7 Breast Cancer: *Semantic Space* for the function importance measure corresponding to the 225 GP models with five functions in the function set (mapping error = 0.036718). Left: Black rods = models with mapping error < 0.032255 (lower error class), gray cones = models with mapping error within [0.032255, 0.038053) (medium error class), light spheres = models with mapping error ≥ 0.038053 (higher error class). Right: Low error class (rods) with semitransparent membranes wrapping the medium and higher error classes for comparison

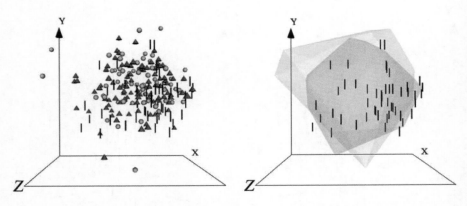

Figure 5.3.8 Breast Cancer: *Semantic Space* for the function importance measure corresponding to the 225 GP models with eight functions in the function set (mapping error = 0.053331). Left: Black rods = models with mapping error < 0.033358 (lower error class), gray cones = models with mapping error within [0.033358, 0.037330) (medium error class), light spheres = models with mapping error ≥ 0.037330 (higher error class). Right: Low error class (rods) with semitransparent membranes wrapping the medium and higher error classes for comparison

the medium and higher error classes are represented. This picture is consistent with rough set results, where the number of decision rules generated was 205, which is almost equal to the number of models analyzed. This indicates lack of generalization capability (mostly are single-case rules) and high data dispersion, as shown by the visual spaces.

Hybrid Variable and Function Spaces: Breast Cancer

Predictor variables and functions from the function set are the constituents of GP models and from a general perspective they can be seen as features describing them. Their individual importance (Equation (5.3.5)) can be arranged in a hybrid matrix composed of predictor variables and functions, amenable to visual exploration in the same way as done for the variables and functions separately. In the Breast Cancer GP models with five functions in the function set, 14 hybrid features (nine variables and five functions) can be used to characterize each model. A representation of the similarity structure of the set of models from this hybrid description perspective is shown in Figure 5.3.9. The combined effect of predictor variables and the functions as model descriptors result in a clearer model distribution in the visual space, where the region with high model density is clearly shifted towards the increasing values of the X axis. This behavior was observed in the previous section, but it is more clearly visible in the hybrid case. The distribution of medium and high error

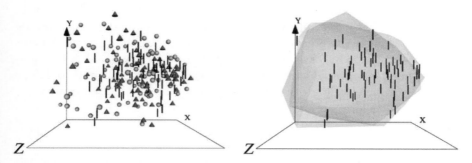

Figure 5.3.9 Breast Cancer: *Semantic Space* for the hybrid variable and function importance measure corresponding to the 225 GP models with five functions in the function set (mapping error = 0.044973). Left: Black rods = models with mapping error < 0.032255 (lower error class), gray cones = models with mapping error within [0.032255, 0.038053) (medium error class), light spheres = models with mapping error ≥ 0.038053 (higher error class). Right: Low error class (rods) with semitransparent membranes wrapping the medium and higher error classes for comparison

models is very similar when both sides of Figure 5.3.9 are compared and the predominance of low error models for higher values of X is clearly shown in Figure 5.3.9(right), indicating that certain combinations of variables and functions are more prone to produce better GP models.

5.3.5 Future Directions

This chapter presented examples of Virtual Reality *Visual Spaces* for the purposes of Visual Data Mining and Meta-Mining. Examples were computed using both classical optimization and Genetic Programming approaches on two medical data sets. All of the visual spaces computed (data and semantic) were unsupervised, but supervised and mixed-paradigm spaces would provide additional valuable information as well. In the same way, the semantic visual spaces presented for GP meta-mining were implicit, preventing any interpretation of the variables of the transformed space in terms of the GP predictor variables and/or functions. An approach using genetic programming (i.e. genetic programming for meta-mining genetic programming results) would have produced explicit analytic equations from which a potential explanation could be offered for the relationship between specific variables and/or functions and model quality.

The use of visual spaces (of different kinds) for data mining and for meta-mining is a valuable tool in computational intelligence and machine learning oriented to the construction of data-driven models. It is an appealing approach complementing statistical and other techniques for model analysis, particularly when exploring large masses of distributed and grid computing experiments. The convergence of these technologies with the developments in virtual reality and 3D systems (ranging from small personal kits for desktop computers to affordable small-scale 3D systems) will contribute to change the way in which exploratory data analysis is conceived and performed.

References

Abraham T and Roddick JF 1999 Incremental meta-mining from large temporal data sets. In Kambayashi Y, Lee DK, Lim E-P, Mohania M, and Masunaga Y (eds) *Advances in Database Technologies: Proceedings of the First International Workshop on Data Warehousing and Data Mining*. Springer-Verlag, Berlin, pp. 41–54.

Barton AJ 2009 Learning the neuron functions within neural networks based on genetic programming. M.C.S. thesis, Ottawa-Carleton Institute for Computer Science, School of Computer Science, Ottawa, Ontario, Canada.

Bazan J, Skowron A, and Synak P 1994 Dynamic reducts as a tool for extracting laws from decision tables. In Proceedings of the 8th International Symposium on Methodologies for Intelligent Systems, Charlotte, NC, USA. *Lecture Notes in Artificial Intelligence* **89**, 346–355.

Belkin M and Niyogi P 2003 Laplacian eigenmaps for dimensionality reduction and data representation. *Neural Computation* **15** (6), 1373–1396.

Bishop CM, Svensén M, and Williams CKI 1998 The generative topographic mapping. *Neural Computing* **10**, 215–234.

Borg I and Lingoes J 1987 *Multidimensional Similarity Structure Analysis*. Springer-Verlag, Berlin.

Chandon JL and Pinson S 1981 *Analyse typologique. Théorie et applications*. Masson, Paris.

De Preter K, De Brouwer S, Van Maerken T, Pattyn F, Schramm A, Eggert A, Vandesompele J, and Speleman F 2009 Meta-mining of neuroblastoma and neuroblast gene expression profiles reveals candidate therapeutic compounds. Clinical Cancer Research. PMID 19435837.

ECJ 2007 A Java-based evolution computing research system. http://www.cs.gmu.edu/~eclab/ projects/ecj/.

Ferreira C 2006 *Gene Expression Programming: Mathematical Modeling by an Artificial Intelligence*. Springer-Verlag, Berlin.

Gower JC 1973 A general coefficient of similarity and some of its properties. *Biometrics* **1** (27), 857–871.

Hartigan JA 1975 *Clustering Algorithms*. John Wiley & Sons, New York.

Hastie T 1984 Principal curves and surfaces. Ph.D., Department of Statistics, Stanford University.

Hastie T and Stuetzle W 1988 Principal curves. *Journal of the American Statistical Association* **84**, 502–516.

Hinton GE and Roweis ST 2003 Stochastic neighbor embedding. *Advances in Neural Information Processing Systems* **15**, 833–840.

Koza J 1989 Hierarchical genetic algorithms operating on populations of computer programs. In *Proceedings of the 11th International Joint Conference on Artificial Intelligence*, Vol. 1, pp. 768–774.

Koza J 1992 *Genetic Programming: On the Programming of Computers by Means of Natural Selection*. MIT Press, Boston, MA.

Koza J 1994 *Genetic Programming II: Automatic Discovery of Reusable Programs*. MIT Press, Boston, MA.

Koza J, Bennett F, Andre D, and Keane M 1999 *Genetic Programming III: Darwinian Invention and Problem Solving*. Morgan Kaufmann, New York.

Kruskal J 1964 Multidimensional scaling by optimizing goodness of fit to a nonmetric hypothesis. *Psichometrika* **29**, 1–27.

Mangasarian OL and Wolberg WH 1990 Cancer diagnosis via linear programming. *SIAM News* **23** (5), 1–18.

Pawlak Z 1991 *Rough Sets: Theoretical aspects of reasoning about data*. Kluwer Academic Publishers, Netherlands.

Press W, Flannery B, Teukolsky S, and Vetterling W 1992 *Numeric Recipes in C*. Cambridge University Press, Cambridge.

Roweis ST and Saul LK 2000 Nonlinear dimensionality reduction by locally linear embedding. *Science* **290** (5500), 2323–2326.

Sammon JW 1969 A non-linear mapping for data structure analysis. *IEEE Transactions on Computers* **C18**, 401–408.

Valdés JJ 2002 Virtual reality representation of relational systems and decision rules. In Hajek P (ed.) *Theory and Application of Relational Structures as Knowledge Instruments*. Meeting of the COST Action, Prague, p. 274.

Valdés JJ 2003 Virtual reality representation of information systems and decision rules. In *Lecture Notes in Artificial Intelligence*, Vol. 2639 of LNAI. Springer-Verlag, Berlin, pp. 615–618.

Valdés J and Barton 2005 Virtual reality visual data mining with nonlinear discriminant neural networks: application to leukemia and Alzheimer gene expression data. Proceedings of the International Joint Conference on Neural Networks (IJCNN), Montreal, Quebec, Canada.

Valdés JJ and Barton AJ 2006 Virtual reality spaces for visual data mining with multi-objective evolutionary optimization: implicit and explicit function representations mixing unsupervised and supervised properties. IEEE Congress of Evolutionary Computation (CEC), Vancouver, pp. 5592–5598.

Valdés JJ, Orchard R, and Barton AJ 2007a Exploring medical data using visual spaces with genetic programming and implicit functional mappings. In GECCO Workshop on Medical Applications of Genetic and Evolutionary Computation. The Genetic and Evolutionary Computation Conference (GECCO), London, UK.

Valdés JJ, Barton AJ, and Orchard R 2007b Virtual reality high dimensional objective spaces for multi-objective optimization: an improved representation. Proceedings of the IEEE Congress on Evolutionary Computation, Singapore.

Whitney EM, Ghaleb AM, Chen X, and Yang VW 2006 Transcriptional profiling of the cell cycle checkpoint gene Krüppel-like factor 4 reveals a global inhibitory function in macromolecular biosynthesis. *Gene Expression* **13** (2), 85–96.

Wikipedia 2007 Cell cycle. Wikipedia, The Free Encyclopedia. http://www.wikipedia.org/.

Wolberg WH and Mangasarian OL 1990 Multisurface method of pattern separation for medical diagnosis applied to breast cytology. In *Proceedings of the National Academy of Sciences USA*, Vol. 87, pp. 9193–9196.

Wróblewski J 2001 Ensembles of classifiers based on approximate reducts. *Fundamenta Informaticae* **47**, 351–360.

Young F 1981 *Introduction to Multidimensional Scaling: Theory, Methods, and Applications*. Academic Press, New York.

6

Advanced Modelling, Diagnosis and Treatment using GEC

6.1

Objective Assessment of Visuo-spatial Ability using Implicit Context Representation Cartesian Genetic Programming

Michael A. Lones and Stephen L. Smith
Department of Electronics, University of York, York, UK

6.1.1 Introduction

Visuo-spatial ability can be defined as a person's manipulation of 'visual representations and their spatial relationships' [5] and is used in many everyday activities such as parking a car, reading a map or pouring a drink. A deficit of visuo-spatial ability is observed in many neuropsychological conditions, such as stroke, Parkinson's disease and Alzheimer's disease, and consequently is an important symptom. However, conventional measurement of visuo-spatial ability can be time-consuming and is often subjective.

Genetic and Evolutionary Computation: Medical Applications   Edited by Stephen L. Smith and Stefano Cagnoni
© 2011 John Wiley & Sons, Ltd

The aim of the work described here is to develop an automated, objective assessment of visuo-spatial ability that can be made easily and reliably. Section 6.1.2 considers the conventional evaluation of visuo-spatial ability using a traditional test environment based on figure copying. Section 6.1.3 gives an overview of Implicit Context Representation Cartesian Genetic Programming and a revised development process used for this research. Section 6.1.4 describes how this algorithm is used to evolve classifiers which automatically classify subjects' drawings. Section 6.1.5 presents results. Conclusions and discussion are presented in Section 6.1.6.

6.1.2 Evaluation of Visuo-spatial Ability

Early diagnosis of any neurological dysfunction is highly desirable as it may permit appropriate therapies or treatment to slow the progression of the disease and minimise its symptoms. However, absolute diagnosis of many neurological conditions is only possible by examining brain tissue and is therefore impractical whilst the patient is alive. Due to this difficulty, determining the presence of a neurological disease is most often a diagnosis of exclusion, where the physician will try to find other causes of the symptoms, often by using laboratory tests and neuroimaging techniques.

An important part of the diagnosis and monitoring of the disease is to perform a neurological examination to evaluate the extent of the impairment of the patient. For example, the most common method of diagnosis based on these examinations for Alzheimer's disease (AD) is the NINCDS-ADRDA Alzheimer's Criteria [13], which examine eight cognitive domains: memory, language, perception, attention, constructive ability, orientation, problem solving and functional ability. Problems within these domains could suggest the onset of AD and the criteria lead to four possible outcomes: definite, probable, possible and unlikely AD.

Geometric shape-drawing tasks are often used to evaluate visuo-spatial neglect. Several tests have been developed, such as the Clock Drawing Test, the Rey–Osterrieth Complex Figure Test and cube-drawing tests. Research into cube-drawing ability has not only shown that it is a useful tool in the detection of AD, but also that it is good at the detection of very mild AD [16]. For cube-drawing assessments, detailed marking criteria are used to grade the cube and hence determine the level of impairment. One example of such criteria is presented in [1], which is used to mark the development of cube-drawing ability of 7 to 10 year olds and shows many similarities with the criteria used in [16] to mark drawings of elderly and AD patients. The scoring system taken from [1] is as follows:

1) A single square or rectangle of any orientation.
2) A set of interconnected squares or rectangles numbering more or less than the number of visible faces in the cube (three) or single trapezoid with some appropriate use of oblique lines.
3) A set of three interconnected squares or rectangles not appropriately arranged to represent the visible arrangement of faces in the cube or a set of interconnected squared or rectangles numbering more or less than the number of visible faces in the cube including some appropriate use of oblique lines.
4) A set of three interconnected squares or rectangles appropriately arranged to represent the visible arrangement of the faces of the cube or an inappropriately arranged set of three outlines including some appropriate use of oblique lines.
5) Drawings that show only visible faces of the cube appropriately arranged (as noted previously) and that reveal crude attempts to show depth through use of oblique lines, curvature, or modification to angles.
6) Drawings that approximate to oblique projection or linear perspective or drawings that approximate well to oblique projection or linear perspective but that are drawn to a horizontal ground line rather than to an oblique ground plane.
7) Drawings that are close approximations to oblique projection or linear perspective but that contain some inaccuracies in angular relations between lines.
8) Accurate portrayals of a cube in oblique projection or linear perspective.

Figure 6.1.1 shows eight example drawings which have been classified based on this system.

Application of the assessment criteria by trained assessors can vary and, hence, is arguably unreliable, so it is desirable to produce an assessment mechanism that will be able to classify cube drawings in a completely objective way. Guest *et al.* [8] implement an algorithm to extract components from static hand-drawn responses for two figure-copying tests and one figure-completion test. First the image is 'skeletonised' then split into its horizontal, vertical and diagonal components by using directional neighbourhood identification. The components are then assessed based on certain features, such as component omissions, length difference and spatial differences in order to

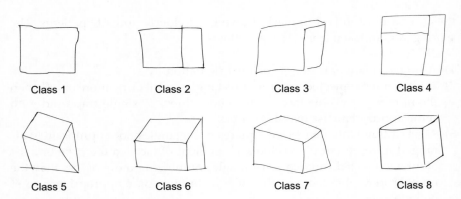

Figure 6.1.1 Eight classifications of cube drawings using the marking system described by Bremner *et al.* [1]

examine the divergence between neglect and control responses. In [7] they extend this idea to include the analysis of dynamic performance features such as pen lifts, movement time and drawing time to improve the sensitivity of the assessment. By looking at these dynamic features they conclude that they can gain an additional understanding of the condition. However, the algorithms described in [7] and [8] use rigid sets of rules designed by the authors based on observed differences. This chapter proposes a method in which this level of subjectivity is removed by using an evolutionary algorithm, Implicit Context Representation Cartesian Genetic Programming, to identify the features used for classifying subject responses.

6.1.3 Implicit Context Representation CGP

Implicit Context Representation Cartesian Genetic Programming (IRCGP) [17] is a form of Cartesian Genetic Programming (CGP) [15] which uses an implicit context representation [10–12].

CGP is a graph-based genetic programming system which has been shown to perform well within a wide range of problem domains. A CGP solution consists of an n-dimensional grid (where n is typically 1 or 2) in which each grid location contains a function. Program inputs and outputs are delivered to and taken from specific grid cells. Interconnections between functions, inputs and outputs are expressed in terms of the grid's Cartesian co-ordinate system. Variation operators (mutation and crossover) are able to alter both the function present within a grid cell and the connections between components.

The efficacy of CGP has been attributed to both implicit reuse of sub-expressions (due to its graphical representation) and its use of functional redundancy [14]. However, CGP programs are positionally dependent, since the behavioural context of a function (i.e. where it receives its inputs from and sends its output to) is dependent upon its Cartesian co-ordinate within the program's representation. Positional dependence, in turn, causes disruptive behaviour during recombination. A consequence of this is that CGP programs, in common with standard GP programs, do not generally respond well to crossover operators [2] (except where the operator is a good match to the problem [4]).

Implicit context is a means of introducing positional independence to GP solution representations. The principle behind implicit context is that interconnections between solution components (outputs, functions and input terminals) are specified in terms of each component's *functional context within the solution* rather than its physical location within the solution. Consequently, when a component's location changes following crossover (or mutation), its expected behaviour will not change. Implicit context representation was originally developed for the Enzyme GP system [12].

In standard CGP, a function's inputs are specified by Cartesian grid references. In IRCGP, a function's inputs are specified by *functionality profiles* which are then resolved to Cartesian grid references during a simple constraint satisfaction development process. Formally, a functionality profile is a vector in an n-dimensional space where each dimension corresponds to a function or terminal. This vector describes the relative occurrence of each function and terminal, weighted by depth, within an expression. In effect, it provides a means of representing and comparing (through vector difference) the functional behaviour of an expression. Details of how functionality profiles are constructed in IRCGP can be found in Smith *et al.* [17].

The functionality profile(s) associated with a function component describes the sub-expression(s) from which it would prefer to receive its input(s). Prior to evaluation, an IRCGP solution undergoes a development process in which each function and output component attempts to find the sub-expression(s) which most closely match its functionality profile(s). In previous work, this has been achieved using the top-down development process of Enzyme GP [11]. However, the grid-structure and feed-forward nature of CGP means that this can also be achieved using a bottom-up process – which, in turn, leads to more accurate matching between functionality profiles and the sub-expressions which they connect to.

The bottom-up IRCGP development process is illustrated in Figure 6.1.2. As in standard CGP, function components (Fn) are ordered from the top left

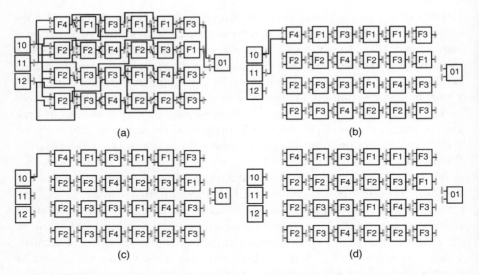

Figure 6.1.2 Bottom-up development process for satisfying functionality profiles

to the bottom right of the grid (in column then row order). To prevent recurrent connections, a component may only receive inputs from the outputs of components lower in the ordering. Also in common with standard CGP, connections are subject to a levels back constraint – meaning that a component may only receive inputs from one located within a specified number of preceding columns.

Starting with an initially unconnected grid (Figure 6.1.2a), development begins at the first input of the first function component (Figure 6.1.2b). Since there are no other function components before this component, it may only choose from the program's inputs (labelled I0–I2) – and will choose whichever one most closely matches its functionality profile (I0 in this case). The development process will then move on to the function's other input(s) (Figure 6.1.2c) and then up through the other function components in the network until all component inputs have been satisfied (Figure 6.1.2d).

6.1.4 Methodology

In previous work on automated Parkinson's diagnosis [18], it was shown that evolved IRCGP solutions are able to describe acceleration patterns which are over-represented in the movements of Parkinson's patients relative to control subjects. In this work, we extend the approach to the multi-class cube-drawing classification task described in Section 6.1.2. In particular, we hypothesise that the development or degradation of visuo-spatial ability is reflected in the physical movements of subjects when carrying out the cube-drawing task and, furthermore, that these patterns of movement (if adequately described) can be used as a basis for automated classification.

6.1.4.1 Data Collection

Drawings made by children can be easily digitised by using a commercial digitising tablet. Use of an inking, wireless pen enables a traditional pen and paper environment to be preserved, reducing stress and distraction in the participants. Modern digitising tablets can sample pen movements up to 200 times per second at a spatial resolution of up to 5000 lines per inch, enabling very fine reproduction of the drawings made.

Drawings were taken from children ranging from 7 to 11 years attending a conventional state school (having obtained local ethical approval and informed consent). Each child was asked to make several attempts at drawing a copy of a cube. Once the data was collected, the cubes were manually classified by two independent markers using the scheme of [1] (see Section 6.1.2). No drawings were identified as class 1.

In total, 120 drawings were recorded from 40 subjects (each providing two to four drawings). The position of the pen, both whilst drawing and whilst lifted from the tablet, was recorded using a Wacom Intuos3 pen tablet at a sampling rate of 200 Hz. This position data was then converted to an acceleration sequence using discrete differentiation, truncated to one standard deviation around the mean (to remove skew) and smoothed using a moving average filter of size 2 (to reduce noise). In order to normalise with respect to drawing time, the resulting sequences were scaled (using linear interpolation if necessary) to a standard length of 4000. Acceleration values were then quantised to the integer range $[-10, 10]$. This quantisation is intended to remove minor fluctuations from the acceleration sequences, since classifiers based on minor fluctuations are likely to be less meaningful and more fragile than those based upon gross acceleration features.

Based on the manual classification of the drawings, the corresponding acceleration sequences were divided into training and test sets, maintaining a ratio of approximately 2:1 both overall and within each class. In order to prevent possible bias, multiple drawings from an individual subject were not split between training and test sets.

6.1.4.2 Evaluation

An acceleration sequence is presented to an evolved IRCGP solution as a sequence of overlapping data windows of length 30. For each of these windows, the IRCGP solution calculates an output. Any value less than zero is interpreted as a positive match. The classification of an acceleration sequence is given by the number of positive matches over all windows.

Receiver Operating Characteristic (ROC) analysis is used to measure an evolved classifier's fitness – its ability to discriminate between data classes. A ROC curve plots true positive rate (TPR) against false positive rate (FPR) across the range of possible classification thresholds, where:

$$\text{TPR} = \frac{\text{Number of positive examples correctly classified}}{\text{Number of positive examples}}$$

$$\text{FPR} = \frac{\text{Number of negative examples incorrectly classified}}{\text{Number of negative examples}} \quad (6.1.1)$$

The area under a ROC curve (known as AUC) is often used as a measure of classifier accuracy, since it is equivalent to the probability that the classifier will rank a randomly chosen positive example higher than a randomly chosen negative example [6]. AUC scores fall within the range [0,1], where 1 indicates perfect discrimination between positive and negative data sets, 0.5 indicates no ability to discriminate, and 0 indicates that negative data is always ranked higher than positive data (i.e. perfect classification can be achieved by inverting the classifier's output).

The AUC metric can be extended to multi-class classifiers by taking the mean of the AUCs of each pair of classes [9], thus measuring the overall pairwise discriminability of the classifier – in effect, how well the classifier separates the classes within its output range. Hand and Till [9] define this metric as:

$$\text{AUC}_{\text{multiclass}} = \frac{2}{|C|(|C| - 1)} \sum_{\{c_i, c_j\} \in C} \text{AUC}(c_i, c_j) \quad (6.1.2)$$

where C is the set of classes and $\text{AUC}(c_i, c_j)$ is the area under the ROC curve when separating classes c_i and c_j. In this work, we do not require that the classifier ranks the classes in their original numerical order, only that it separates the classes within the output range. This is done by inversely mapping pairwise AUC scores in the range [0,0.5) to the range (0.5,1.0].

6.1.4.3 Parameter Settings

We carried out five runs of 200 generations using a population of 200 classifiers. Child solutions were generated using uniform crossover and mutation

Table 6.1.1 Function set

Function	Symbol	Description
Add	+	Returns the sum of its two inputs
Subtract	−	Returns the difference of its two inputs
Mean	M	Returns the mean of its two inputs
Min	<	Returns the lesser of its two inputs
Max	>	Returns the greater of its two inputs
Absolute	||	Returns the absolute value of its input
Negate	!	Returns its input multiplied by −1

Table 6.1.2 Pairwise AUC scores for best evolved classifier upon training and test sets. High scores (0.2 >= AUC >= 0.8) are shown in bold face. The last column indicates whether training and test scores are correlated

Pair	Train AUC	Test AUC	Correlated?
2/3	**0.83**	**1.00**	yes
2/4	0.74	**0.80**	yes
2/5	**0.05**	**1.00**	
2/6	0.66	**0.90**	yes
2/7	**0.94**	**1.00**	yes
2/8	**0.89**	**0.90**	yes
3/4	0.69	0.60	yes
3/5	**0.00**	**1.00**	
3/6	0.39	**0.80**	
3/7	**0.87**	**1.00**	yes
3/8	**0.85**	**0.80**	yes
4/5	**0.10**	**0.80**	
4/6	0.28	0.68	
4/7	0.57	0.70	yes
4/8	0.46	0.64	
5/6	**0.86**	0.38	
5/7	**0.99**	0.21	
5/8	**0.96**	**0.13**	
6/7	**0.85**	0.44	
6/8	0.78	0.33	
7/8	0.33	0.45	yes

in equal proportion. The mutation rate was 6% for functions and 3% for functionality dimensions. We used a CGP grid size of 9 rows by 8 columns. These values were determined experimentally. The function set is defined in Table 6.1.1.

6.1.5 Results

The best evolved classifier had a multi-class AUC score of 0.80 on both the training and test sets. Table 6.1.2 shows its AUC scores for each pair of classes. Of the 21 pairs of classes, 12 have high AUC scores for both the training and test sets. However, in some cases these do not correlate across the training and test sets, which may indicate over-learning. Nevertheless, the lower-numbered classes are fairly well separated from the higher-numbered classes, suggesting that the evolutionary algorithm has found a meaningful pattern.

Figure 6.1.3 shows the best matching sequence windows against three of the highest scoring classifiers. A prominent feature in these patterns is the presence of a dual deceleration peak. This can also be seen in Figure 6.1.4, which overlays all the sequence windows that were matched positively by the best of these classifiers. Whilst there is a fair amount of variance between the matches, most exhibit a dual peak and a region of relatively constant acceleration either side. This is especially the case for the stronger matches (i.e. those which receive higher classification scores), indicated by thicker lines in the diagram.

Figure 6.1.5 shows the locations of these match windows within four example drawings from different classes. It can be seen that whilst matches occur throughout the drawing in the lower-numbered classes, they tend to cluster around corners in the higher-numbered classes. This leads to the hypothesis that the evolved expression is recognizing regions in which the subject hesitates or carries out jittery motion. In the case of lower-numbered classes, this may reflect the subject's general unfamiliarity with the shape they are attempting to draw; whereas in higher-numbered classes, uneven motion only occurs around changes in direction.

Figure 6.1.6 shows the pattern-matching expression used by the best classifier. Whilst it is difficult to interpret the exact behaviour of the classifier from this expression, it can be seen that the sub-expression '18-17' is re-used four times when calculating a match. Offsets 17 and 18 correspond to the first peak in Figure 6.1.4, suggesting that the rate of change of acceleration at this point is an important feature underlying classification. It is also notable that the classifier considers multiple pairs of offsets (7/21, 8/19, 13/22, and 10 with 17/18), where one occurs in the region preceding the acceleration peaks and the other occurs within the peaks. This strengthens the idea that

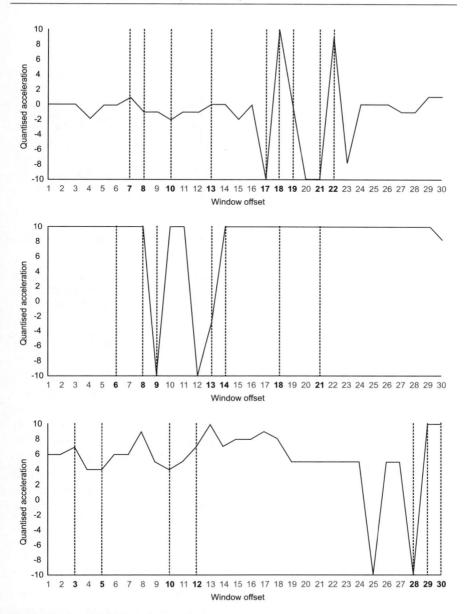

Figure 6.1.3 Best matching acceleration sequence windows against (top) the best and two other high scoring classifiers. Vertical dashed lines show the window offsets used by the evolved expression

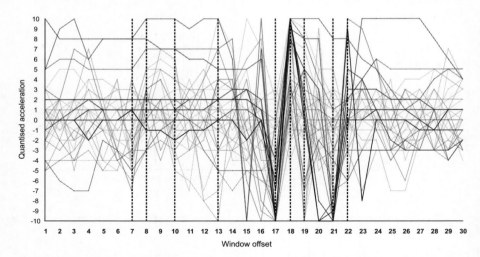

Figure 6.1.4   Overlay of all positive matches to the best evolved classifier across all sequences. The weight of a line indicates the strength of the corresponding match. Vertical dashed lines show the window offsets used by the evolved expression

the classifier is looking for two different features, i.e. a region of relatively constant acceleration followed by acceleration peaks.

The different separations between the pairs of offsets (14, 11, 9 and 7-8, respectively) also suggests that the classifier is looking for the features at different time scales within the match window. The use of fixed offsets is one of the limitations of the current classifier model, since it requires relatively complex expressions in order to describe a feature occurring at different time scales. Whilst normalisation of sequence length mitigates this effect at a gross level, it is still likely that there will be timing variations between the responses of different subjects within a class. In future work, we plan to address this issue by looking at the utility of feature-based encodings, such as time domain signal coding [3], which are less sensitive to scale.

6.1.6 Conclusions

In this work, we have demonstrated how Implicit Context Representation Cartesian Genetic Programming can be used to identify meaningful patterns of subject movement within recordings of children carrying out cube-drawing tasks. Our results suggest that the resulting classifiers can be used to categorise a child's relative stage of neurological development. In future work,

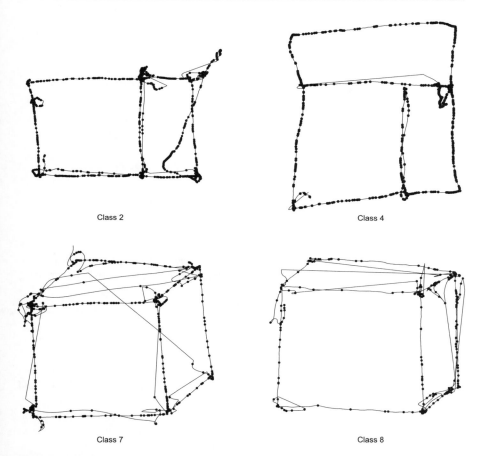

Class 2   Class 4

Class 7   Class 8

Figure 6.1.5 Examples of drawings from four different classes, showing match locations against the best evolved classifier. Filled circles indicate the start of positively matched windows within the drawing's corresponding acceleration sequence

we hope to apply this technique to the automated diagnosis of neurological diseases such as Parkinson's and Alzheimer's.

In addition, this work has shown how Implicit Context Representation Cartesian Genetic Programming can be applied to a multi-class classification problem by using a multi-class ROC analysis-based fitness function. We have also introduced a bottom-up development process to the algorithm. Our practical experience suggests that this improves the performance of the method, and in future work, we hope to expose it to a more in-depth analysis.

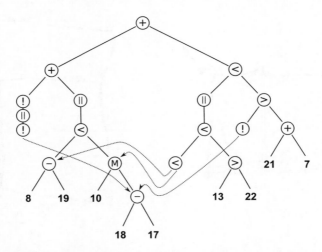

Figure 6.1.6 Best evolved classifier. Dotted lines indicate implicit reuse of sub-expressions within the IRCGP network. Numbers indicate offsets in the matching window

References

[1] J. G. Bremner, R. Morse, S. Hughes and G. Andreasen. Relations between drawing cubes and copying line diagrams of cubes in 7- to 10-year-old children. *Child Development* **71**(3):621–634, 2000.

[2] X. Cai, S. L. Smith and A. M. Tyrrell. Positional independence and recombination in Cartesian genetic programming. *Proceedings of the 2006 European Conference on Genetic Programming* (EuroGP), pp. 351–360, 2006.

[3] E. D. Chesmore. Application of time domain signal coding and artificial neural networks to passive acoustical identification of animals. *Journal of Applied Acoustics* **62**:1359–1374, 2001.

[4] J. Clegg, J. A. Walker and J. F. Miller. A new crossover technique for Cartesian genetic programming. Genetic and Evolutionary Computation Conference (GECCO 2007), 2007.

[5] W. Dorland (ed.). *Dorland's Medical Dictionary for Health Consumers*. Saunders (an imprint of Elsevier), 2007.

[6] T. Fawcett. An introduction to ROC analysis. *Pattern Recognition Letters* **27**:861–874, 2006.

[7] R. Guest and M. Fairhurst. A novel multi-stage approach to the detection of visuo-spatial neglect based on the analysis of figure-copying tasks. *Proceedings of the Fifth International ACM Conference on Assistive Technologies*, pp. 157–161, 2002.

[8] R. Guest, M. Fairhurst and J. Potter. Automated extraction of image segments from clinically diagnostic hand-drawn geometric shapes. *Proceedings of the 26th Euromicro Conference*, Vol. 2, pp. 440–446, 2000.

[9] D. J. Hand and R. J. Till. A simple generalization of the area under the ROC curve to multiple class classification problems. *Machine Learning* **45**(2):171–186, 2001.

[10] M. A. Lones. Enzyme Genetic Programming: Modelling biological evolvability in genetic programming. Ph.D. thesis, Department of Electronics, University of York, 2003.

[11] M. A. Lones and A. M. Tyrrell. Biomimetic representation with enzyme genetic programming. *Genetic Programming and Evolvable Machines* **3**(2):193–217, 2002.

[12] M. A. Lones and A. M. Tyrrell. Modelling biological evolvability: implicit context and variation filtering in enzyme genetic programming. *BioSystems* **76**(1–3): 229–238, 2004.

[13] G. McKhann, D. Drachman, M. Folstein, R. Katzman, D. Price and E. M. Stadlan. Clinical diagnosis of Alzheimer's disease: report of the NINCDS-ADRDA Work Group under the auspices of the Department of Health and Human Services Task Force on Alzheimer's Disease. *Neurology* **34**:93–94, 1984.

[14] J. F. Miller and S. L. Smith. Redundancy and computational efficiency in Cartesian genetic programming. *IEEE Transactions on Evolutionary Computation* **10**(2):167–174, 2006.

[15] J. F. Miller and P. Thomson. Cartesian genetic programming. In R. Poli, W. Banzhaf, W. B. Langdon, J. F. Miller, P. Nordin and T. C. Fogarty (eds), *Third European Conference on Genetic Programming*, Vol. 1802 of *Lecture Notes in Computer Science*, 2000.

[16] Y. Shimada, K. Meguro, M. Kasai, M. Shimada, H. Ishii, S. Yamaguchi and A. Ya-madori. Necker cube copying ability in normal elderly and Alzheimer's disease. A community-based study: the Tajiri project. *Psychogeriatrics* **6**(1):4–9, 2006.

[17] S. L. Smith, S. Leggett and A. M. Tyrrell. An implicit context representation for evolving image processing filters. *Proceedings of the 7th Workshop on Evolutionary Computation in Image Analysis and Signal Processing*, Vol. 3449 of *Lecture Notes in Computer Science*, pp. 407–416, 2005.

[18] S. L. Smith, P. Gaughan, D. M. Halliday, Q. Ju, N. M. Aly and J. R. Playfer. Diagnosis of Parkinson's disease using evolutionary algorithms. *Genetic Programming and Evolvable Machines* **8**(4):433–447, 2007.

6.2

Towards an Alternative to Magnetic Resonance Imaging for Vocal Tract Shape Measurement using the Principles of Evolution

David M. Howard, Andy M. Tyrrell, and Crispin Cooper
Department of Electronics, University of York, York, UK

6.2.1 Introduction

Electronic voice synthesis has a number of applications today, and it is perfectly possible for it to produce a highly *intelligible* speech output. However, its output is rarely, if ever, mistaken for the voice of a human; that is, it rarely sounds *natural*. Sorting out and understanding which elements of human speech production contribute to our perception of naturalness is a key goal of today's speech research.

Genetic and Evolutionary Computation: Medical Applications   Edited by Stephen L. Smith and Stefano Cagnoni
© 2011 John Wiley & Sons, Ltd

Physical modelling synthesis techniques have been used successfully for electronic music synthesis [1], where outputs are often described as being *natural sounding*, or *organic*. This contrasts with outputs obtained from more traditional music synthesis methods such as additive, subtractive, wavetable or sampling synthesis [e.g. 2], which are often described as being *cold* or *lifeless* by players and audience alike. In addition, listeners find that the more traditional music synthesis methods become less interesting with extended exposure [3]. If physical modelling synthesis can enhance the naturalness of synthesis in the musical domain, it seems very plausible to apply it to the voice synthesis arena.

Electronically synthesized speech first started to become practicably viable in the late 1930s. Since then, a number of different approaches have been adopted which range between modifying natural speech recordings at one extreme to direct modelling of the physiological processes involved in speech production at the other. A useful overview of speech synthesis methods is given by Styger and Keller [4] under the following four categories.

1. Manipulation of natural speech waveforms
   no knowledge of speech production mechanism
2. Linear predictive synthesis
   all-pole acoustic model of the vocal tract
3. Formant synthesis
   formant (acoustic resonance) parameters can be varied
4. Vocal tract analogue and articulatory physical modelling
   control of physical parameters relating to vocal tract itself

Styger and Keller ranked these four approaches in this way to allow the application of two scales: (A) the *flow of control parameters* (taken as the sampling rate required to update the synthesis system) and (B) *model complexity* (the knowledge required relating to the speech production process). They note that for (1) (manipulation of natural speech waveforms), the *flow of control parameters* occurs at a maximum rate of once per fundamental period, but it occurs at a minimum rate for (4) (vocal tract analogue and articulatory physical modelling), once per articulatory gesture. Conversely, the *model complexity* in terms of the data necessary to implement each system is the converse; a maximum for (4) and a minimum for (1). Voice synthesis involves a trade-off between the flow of control parameters and model complexity, and in order to create a highly natural result that is generic to the synthesis of the voice of any speaker, (4) is the most appropriate approach, involving direct modelling of the human vocal tract and detailed knowledge of vocal tract articulation.

Whilst Holmes has shown that formant synthesis can produce an output which is indistinguishable from the natural original [5], this was only achieved following a painstaking synthesis-by-analysis approach over many months. For his work, detailed comparisons were made between time–frequency–amplitude spectrograms [6] of the original and synthesized versions to enable changes to be made to the synthesis parameters to increase the acoustic similarity between them. Although this does demonstrate that similarity can be obtained, the application of the source/filter model [7] is compromised because spectral changes that are in reality due to the source are compensated for by changes made to settings of the filter. This offers no gain in terms of a better understanding of what is needed to improve naturalness in voice synthesis. In addition, it has more recently been demonstrated that there are important non-linear interactions between the source and the filter that the original source/filter model does not take into account, and it is likely that these make an essential contribution to the naturalness of the perceived output [8].

A move from a formant synthesis technique seems to be appropriate, and the Styger and Keller [4] description suggests that an articulatory based system (their 4 above) is likely to be the best choice. Physical modelling provides a mechanism for articulatory synthesis as well as one that can produce natural sounding results in the musical context. Effects such as non-linear source–tract interaction will emerge as a consequence of the application of the physical modelling process itself. Physical modelling requires that the physical size and shape of the system (here the oral tract) can be specified accurately. Such measurements are usually gained from Functional Magnetic Resonance Imaging (fMRI) of the vocal tract [9].

However, fMRI data acquisition is hampered by a number of practical factors. The fMRI environment is very acoustically noisy, which means that the acoustic feedback paths to the ears of the subject are compromised. Spoken or sung sounds produced whilst lying in the machine are rarely representative of the subject's natural output. The subject has to lie supine in the machine, which is not a normal speaking position from the point of view of posture, breathing or vocal tract soft tissue orientation due to gravitational forces acting perpendicularly to normal. It also takes a considerable time (in speech production terms) to acquire an fMRI image; and a subject has to hold motionless an articulatory gesture for some seconds. Voice data gathered in an fMRI experiment is therefore potentially compromised in terms of the degree to which it represents accurately normal vocal behaviour, and an alternative method would be potentially beneficial.

This is the key focus of this chapter and it is one that we believe is entirely novel.

6.2.2 Oral Tract Shape Evolution

The principles of evolution offer a new computational paradigm for finding oral tract cross-sectional areas and thereby a direct alternative to placing subjects in an fMRI machine. Oral tract shapes are evolved and tested using the two-dimensional digital waveguide mesh (DWM) physical modelling synthesis techniques of Mullen *et al.* [10, 11]. They have demonstrated successful synthesis of stationary vowels (e.g. monophthongs such as the vowels in "rat", "writ" and "rut") as well as vowels where the shape of the oral tract changes dynamically (e.g. diphthongs such as the vowels in "right", "rear" and "rate") by means of an *impedance contour map* [12] along the length of the DWM to enable the mesh shape to be changed without the need to remove or replace mesh elements during synthesis, which ensures that there are no audible discontinuities.

Vocal tract areas used are derived from fMRI images of the oral tract of an adult male [9]. Figure 6.2.1 shows two-dimensional oral tract area waveguide mesh element representations for the vowels in *beat* ("ee"), *Bart* ("ah") and *booed* ("uu"). The differences in oral tract area shapes are clearly visible. The commonly used LF model [13], implemented as a wavetable oscillator to enable its f0 value to be readily altered to model pitch changes, is input at

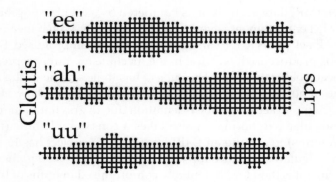

Figure 6.2.1 Two-dimensional oral tract area waveguide mesh element representations of the vowels in *beat* ("ee"), *Bart* ("ah") and *booed* ("uu") derived from fMRI data for an adult male

the glottis end of the mesh to provide an appropriate excitation signal for voiced vowels. The output is summed from the elements at the lip end of the mesh.

When synthesizing steady oral vowels, the mesh shape remains constant. For continuous speech synthesis, however, the shape of the mesh has to be varied dynamically in order to enable the synthesis of sounds such as diphthongs (e.g. the words *eye* or *ear*) for which the oral tract area shape changes dynamically.

There is a method for calculating the shape of the oral tract using an extension of linear predictive coding (LPC) [14]. Rossiter *et al.* [15] implemented a real-time oral tract area display based on LPC analysis as part of their ALBERT system [16] for real-time visual feedback for training professional singers. More recently, this display has been incorporated into the WinSingad PC Windows-based real-time singing training system [17, 18]. There are, however, two key issues relating to the use of LPC analysis for oral tract shape calculation: (1) it assumes an all-pole vocal tract acoustic frequency response which is not always true, and (2) it can produce oral tract shapes that are not necessarily unique since more than one oral tract tube shape can produce a given sound [19].

The main aim of this work is to establish whether or not an evolutionary computation technique can successfully evolve oral tract shapes that produce plausible-sounding vowels when compared to the natural originals. If successful, it will offer immediate improvements over LPC-based oral tract area estimation as well as a potential alternative to fMRI that is both simple and highly cost effective.

6.2.3 Recording the Target Vowels

In order to evolve oral tract shapes, target vowels and some form of synchronously recorded input signal are required for each speaker and each vowel. It is not possible in practice to gain direct access to the glottal pressure wave during the production of a vowel, and some other means is needed to record a suitable input signal. Here, we are dealing with voiced vowels; that is, vowels in which the vibrating vocal folds in the larynx provide the acoustic excitation to the oral tract. The electrolaryngograph [20] enables vocal fold vibration to be monitored non-invasively, and it provides an output waveform (Lx) that relates directly to the nature of the vibrating vocal folds. Although the Lx waveform is not a representation of either the glottal flow or glottal pressure waveform, it is an audio waveform that is directly related to the source of voiced sounds, vocal fold vibration, and it has a characteristic waveshape that is not too dissimilar to the glottal flow waveform.

The vowels were recorded in the acoustically semi-anechoic chamber in the Department of Electronics at the University of York, UK. Two male and two female adult subjects were recorded producing the eight vowels in the words: *boot, beat, bet, Bart, bat, but, Bert* and *bought*. Each vowel was produced thrice, using a rising, falling and flat pitch contour. The Lx waveform from the electrolaryngograph was recorded simultaneously with the speech pressure waveform from a Sennheiser MKH-20 omni-directional microphone and an RME quad microphone amplifier. These two audio channels were recorded in stereo using an Edirol R4 hard disk recorder in linear PCM *.wav* format at 16 bits resolution and a sampling rate of 44.1 kHz. The *.wav* data were transferred digitally to a PC computer for processing. The microphone recordings of the vowels provided the targets during the evolution process and the simultaneously recorded Lx waveforms provided the associated inputs for the physical modelling synthesis process.

It should be noted that Lx waveform is not really appropriate as an input waveform for speech sound production, since it is not directly related to the glottal pressure waveform. However, it does contain many of the natural characteristics of the excitation for voiced speech, and it is very easy to obtain compared with any methodology that is available for measuring glottal pressure, such as inverse filtering. If it works as an input waveform in the evolution process and a satisfactory output waveform is produced, then it will indicate additionally something of the power of the evolution technique in terms of how it is able to take account of such differences.

6.2.4 Evolving Oral Tract Shapes

Bio-inspired computing enables techniques such as genetic evolution to be employed as a computational tool in a number of application areas that involve design and optimization [21–25]. A genetic evolution computational technique requires that there is some way of testing a result from a particular member of a generation, and this is usually achieved by means of deriving an output based on the application of an appropriate input.

In the case of evolving oral tract shapes, an input and target output waveform are required, which in this case are the Lx waveform and vowel speech pressure waveform respectively, and the input is applied to a physical modelling synthesis model, the output from which can be compared to the natural

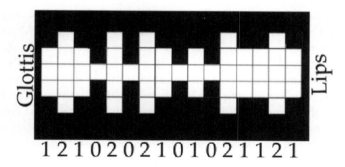

Figure 6.2.2 Oral tract genome in which each digit indicates how many elements there are either side of the midline, around which the mesh is symmetric. In this example, the genome is: 1210202101021121

target waveform. Further details in relation to the methodology can be found in [21] and its application for sung sounds in [25].

In order that oral tract shapes can be evolved, a definition in the form of a genome is required; Figure 6.2.2 shows how the genome of the oral tract is defined for an arbitrary oral tract shape. The genome itself indicates the number of mesh elements on either side of the midline at that point. One element must remain at each point to enable the acoustic pressure to propagate along the tract – we are not dealing with complete constrictions of the oral tract in this work. The example shown in Figure 6.2.1 therefore has the unique genome: 1210202101021121. Since this work is only involved with oral vowels, the tract can never be fully constricted, so there is: (1) a minimum of one element at every position along the mesh, and (2) an odd number of elements across the mesh at all positions.

The oral tract has a length of 16 mesh elements and a minimum and maximum width of 1 and 5 elements, respectively, as illustrated in Figure 6.2.2. The initial population of genotypes or individuals is established by setting up 50 oral tract shapes with randomly shaped oral tract waveguide meshes. The Lx waveform is applied at the glottis end and the output monitored at the lip end and compared with the target natural original in order to evaluate the genotypes for their fitness as a solution. The fitness evaluation is based on comparing the difference between the amplitude frequency spectrum of a 50 ms window taken during the steady-state portion of the target vowel and the output from the mesh.

The 10 (20%) genotypes that have the closest spectral match are deemed to be the fittest and are then copied to the next generation, where they are used as the basis for offspring creation. The remaining 40 (80%) genotypes are discarded and thereby excluded from the next generation. Mutation and crossover operators are defined which operate on one or two genotypes, respectively, to create new members of the new generation from the retained 10 fittest genotypes. The process is iterated until the population converges to a solution, which is established by fitness results that remain stable when compared to the target. In this case, the evolution process was run over 50 generations, and it was repeated twice with a new random set of starter genotypes each time, giving a total of three runs for each vowel.

The target vowels selected for the experiment were those uttered with flat intonation contour, since then the section selected for fitness evaluation would have one less degree of change (pitch) associated with it. A comparison was carried out with vowels spoken on a rising intonation contour and the results were very similar. Since this was a new application for evolutionary computation techniques, various forms of modifications were tried to attempt to improve the overall results. This is a form of human intervention which was based on informal listening to the final outputs to decide whether or not they would pass phonetically as the target vowel. The results of this human intervention by informal experimenter listening are given in Table 6.2.1 using a three-point scale as follows to indicate the extent to which the resulting vowels from the three evolution runs pass phonetically as the target vowels: 1, all three pass; 2, one out of three pass; 3, none pass.

Table 6.2.1 Results for the two male speakers (M1, M2) for the eight vowels uttered on a flat intonation contour. Numerical data indicate how the results over three evolution runs were perceived in terms of passing phonetically as the target vowel during the human process (1, all three runs pass; 2, one out of three runs pass; 3, no runs pass). The effect of modifications (A) and (B) is also shown where there was an observed improvement (A+, B+) or detriment (A−, B−) (modification B was only applied to vowels *boot*, *beat* and *bought*)

Vowel	M1			M2	
boot	3			3	A−,B−
beat	3			3	B+
bat	1			2	A+
but	1			2	A+
bet	2	A−		2	A−
Bert	1			3	A+
Bart	3	A−		1	A−
bought	1	B−		3	B−

Two modifications (denoted as A and B) were implemented in an attempt to improve on these results.

Modification A was originally implemented as a way of speeding up the fitness evaluation process by halving the portion of the target signal that was used (25 ms rather than 50 ms). It turned out that some of the evolved vowels for one subject (M2) were improved (shown by A+ entries in Table 6.2.1). The improvement was not universal though. There were none for subject M1, and some of the resulting vowels were worsened (shown as A− entries in Table 6.2.1).

Modification B was inspired by the observation that the results for those vowels which are produced with a narrow articulation of the tongue with the palate (particularly *beat* and *boot*, but also *bought*), were rated "3" for both subjects (see Table 6.2.1). It was hypothesized that this was due to the low spatial resolution of the oral tract model. A four fold increase in mesh resolution was implemented, increasing the search space by approximately one million, but a good solution was not then achievable and initial evolution runs failed to produce any scores of "1" or "2". Then the width of the mesh at the 10 predefined points along its length was limited to 1, 5, 11 or 17 mesh nodes, corresponding to 2.8 mm, 13.8 mm, 30.3 mm and 46.8 mm, respectively. This was only employed for *boot*, *beat* and *bought*, and the results are shown in Table 6.2.1 as "B+" or "B−" where a difference was observed. One vowel (*beat*) for subject M2 was improved, and the vowel in *bought* was worsened for both subjects.

6.2.5 Results

Results are presented in two forms: oral tract area shapes and spectral comparisons between the evolved vowels and the targets. To guide the comparisons, a listening test was carried out [26] for the evolved vowels of the two male subjects and the results are listed in Table 6.2.2. In this listening test, 12 subjects were asked to decide whether each of the evolved vowels would pass phonetically as its original target, thereby providing an indication as to how close the vowels were in perceptual terms. The results are averaged across all 12 listeners, and they can be compared to the oral tract shape and spectral data presented below.

The listening test results confirm the scores given during the human intervention process (see Table 6.2.1), with high listening test results appearing for vowels which gain a human intervener score of "1" or "2", and low scores

Table 6.2.2 Average responses (%) from 12 listeners who were asked whether each evolved vowel would pass phonetically as the target original for the two male speakers (M1, M2)

Vowel	M1 (%)	M2 (%)
boot	0.0	7.7
beat	23.1	7.7
bat	100.0	100.0
but	100.0	76.9
bet	61.5	92.3
Bert	100.0	53.8
Bart	76.9	92.3
bought	100.0	15.4

for those vowels with a "3". It is hypothesized that evolved vowels with high scores should exhibit spectra that are close to those for their target originals.

Oral tract areas resulting from the evolution process followed by long-term average spectra for the evolved and target vowels are presented in the next sections.

6.2.5.1 Oral Tract Areas

Figure 6.2.3 shows plots of the evolved oral tract areas for both subjects plotted in terms of mesh width genome value against distance from glottis to lips. The glottis and lips are at the left- and right-hand sides of the figure, respectively. The results for both subjects are plotted together for each vowel to enable direct comparison. It is not possible to offer a direct comparison with fMRI data, since the highly acoustical noise levels associated with fMRI machines makes it impossible to produce a useful audio recording.

What the plots do serve to provide is an indication of consistency between the results for the two speakers. It is reasonable to expect that the oral tract shapes for two male speakers should be essentially similar; the tongue will adopt a similar position for all speakers [8].

The oral tract shapes that exhibit the closest matches in terms of a general similarity in shape between the two male speakers are for the vowels in *bat*, *but*, *bet*, *Bert* and *Bart*. The other three (*boot*, *beat*, *bought*) exhibit greater differences. Of particular importance are the relative positions of constrictions in the tract, since these serve to move the formants around [27], and these results suggest that *boot*, *beat*, *bought* may have differences between their formants for these two speakers.

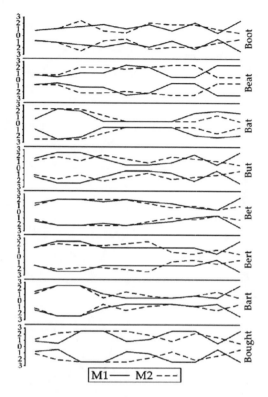

Figure 6.2.3   Evolved oral tract shapes for both male subjects (M1, M2) for the eight vowels (glottis on the left, lips on the right). The y-axis is calibrated in mesh elements

6.2.5.2 Spectral Comparisons

In order to gain an impression of the success of this technique for some of the vowels, long-term average spectra (LTAS) are presented and considered. Since the vowels themselves were produced in isolation, the LTAS will be very similar in shape to the short-term spectrum which was used as the basis for the fitness function evaluation. In each case, the LTAS plotted is taken from the best evolution run, either the original or modification A or B as indicated in Table 6.2.1. The LTAS for the subjects M1 and M2, respectively, for each of the eight vowels are shown in Figures 6.2.4 and 6.2.5 where LTAS plots for the target original and evolved vowels are plotted together to enable direct comparison.

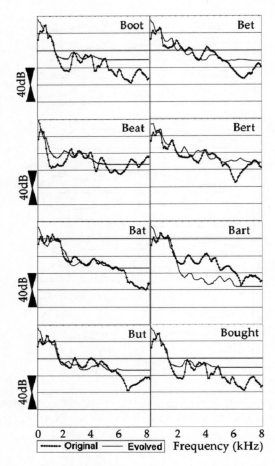

Figure 6.2.4 Long-term average spectra of the target original and evolved vowels for subject M1

Observation of Figure 6.2.4 for speaker M1 indicates that the spectral match is particularly good for the vowels in *bat* and *but*, and the formants are very well matched for *bought*. These scored 100% in the listening test along with *Bert*, for which the overall match is not as clear but the lower formant peaks are well aligned (the lowest three formants are the most crucial for vowel identification [e.g. 27]).

The vowels with the narrowest constriction, *boot* and *beat*, are lacking spectral detail, and neither has evidence of the lowest formant peak (F1). This

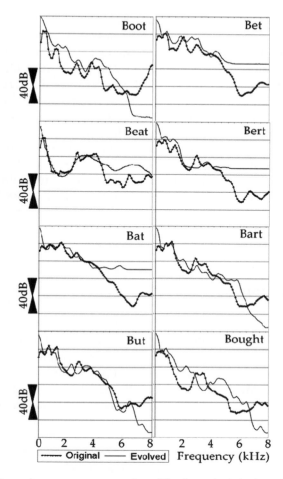

Figure 6.2.5   Long-term average spectra of the target original and evolved vowels for subject M2

will make identifying them difficult, as evidenced by the listening test results (see Table 6.2.2). The vowel in *bet* exhibits peaks that are not apparent in the original, but the lowest two formants are quite well matched, so there will be some perceptual evidence of its phonetic origin as indicated in the listening test results. The remaining vowel in *Bart* is deficient in the overall matching of spectral trend by nearly 40 dB in places, which will somewhat hamper its phonetic identity, although its formant peaks are essentially well matched.

The results for speaker M2 are shown in Figure 6.2.5. Here, the vowels in *bat*, *bet* and *Bart* scored well in the listening test (see Table 6.2.2), and each of these exhibits a good spectral match to the lower formant peaks, especially F1 and F2. The vowel in *but* did less well, and this is most likely due to the peak visible in the LTAS for the evolved version around 2 kHz, which is not apparent for the original. The other four vowels are deficient in their formant peaks, and they score poorly in the listening test. Of particular note is the fact that *boot* and *beat* show no evidence of a first formant peak (the same was true for M1), so once again the vowels with the narrowest constrictions are poorly matched to their targets.

It is worth noting that some of the spectra lack detail in the high-frequency region and this is highly likely to be due to the nature of the Lx waveform used for their excitation. As indicated above, the Lx waveform relates to vocal fold closure and opening and it is not directly linked to the glottal pressure waveform which is the true excitation waveform.

6.2.6 Conclusions

This experiment has demonstrated that it is possible to evolve oral tract shapes using bio-inspired computation techniques, and that repeatable solutions can be achieved. Physical modelling using a digital waveguide mesh provides an appropriate engine for the technique, and it has the advantage that it is directly related to the shape of the oral tract itself. The results indicate that there are issues with respect to evolving oral tract shapes for vowels that require a narrow oral tract constriction, the so-called phonetically *close* vowels, particularly those in *boot* and *beat*.

Modifications were made in an attempt to improve the evolved results, and whilst there was evidence of improvements for some vowels, there are other vowels for which the results are worsened. The modifications employed therefore do not offer a universal solution in terms of improving the results, but it might be that some of their advantages could be made use of in future implementations. It may be that multiple methods could be employed, leaving the fitness evaluation to select the closest result. There is plenty of scope for future work.

The technique itself has the potential to offer a non-invasive method for finding oral tract shapes that would obviate the use of either: (a) fMRI, which is acoustically noisy, involving a supine position of the subject who has to hold a vocal tract posture for a number of seconds, or (b) LPC analysis, which has the potential to produce more than one solution for a given speech input that cannot as yet be constrained in a way appropriate to oral tract articulation.

The fact that solutions were evolved even though the excitation (Lx) was not fully equivalent to the natural glottal excitation during speech is, we believe, quite remarkable. The potential for further useful results being obtained using this technique is, we believe, most promising.

Acknowledgments

The authors thank the speakers and listeners for taking part in the experiments. This work is funded by a UK Engineering and Physical Sciences Research Council (EPSRC) postgraduate studentship, the Department of Electronics at the University of York, and the Future and Emerging Technologies programme (IST-FET) of the European Community, under grant IST-2000-28027 (POETIC). The information provided is the sole responsibility of the authors and does not reflect the Community's opinion. The Community is not responsible for any use that might be made of data appearing in this publication. The Swiss participants in the POETIC project are supported under grant 00.0529-1 by the Swiss government.

References

[1] Pearson, M.D. and Howard, D.M. Recent developments with TAO physical modelling system. Proceedings of the International Computer Music Conference, ICMC-96, 1996, pp. 97–99.

[2] Dodge, C. and Jerse, T.A. *Computer Music: Synthesis, composition and performance*. Schirmer Books, New York, 1985.

[3] Howard, D.M. and Rimell, S. Real-time gesture-controlled physical modelling music synthesis with tactile feedback, EURASIP. *Journal of Applied Signal Processing* 2004;7(15):1001–1006.

[4] Styger, T. and Keller, E. Formant synthesis. In *Fundamentals of Speech Synthesis and Speech Recognition*, E. Keller (ed.). John Wiley & Sons, Chichester, UK, 1994, pp. 109–128

[5] Holmes, J. Synthesis of natural-sounding speech using a formant synthesiser. In *Frontiers of Speech Communications Research*, B. Lindblom and S. Oman (eds). Academic Press, London, 1979, pp. 275–285.

[6] Howard, D.M. Practical voice measurement. In *The Voice Clinic Handbook*, T. Harris, S. Harris, J.S. Rubin and D.M. Howard (eds). Whurr Publishing Company, London, 1998.

[7] Fant, G. *The Acoustic Theory of Speech Production*. Mouton, The Hague, 1960.

[8] Titze, I.R. Theory of glottal airflow and source–filter interaction in speaking and singing. *Acta Acoustica* 2004;**90**(4):641–648.

[9] Story, B.H., Titze, I.R. and Hoffman, E.A. Vocal tract area functions from magnetic resonance imaging. *Journal of the Acoustical Society of America* 1996;**100**(1):537–554.

[10] Mullen, J., Howard, D.M. and Murphy, D.T. Digital waveguide mesh modelling of the vocal tract acoustics. IEEE Workshop on Applications of Signal Processing to Audio and Acoustics, WASPAA, 2003, pp. 119–122.

[11] Mullen, J., Howard, D.M. and Murphy, D.T. Waveguide physical modelling of vocal tract acoustics: improved formant bandwidth control from increased model dimensionality. *IEEE Transactions on Speech and Audio Processing* 2006;**14**(3):964–971.

[12] Mullen, J., Howard, D.M. and Murphy, D.T. Real-time dynamic articulations in the 2D waveguide mesh vocal tract model. *IEEE Transactions on Speech and Audio Processing* 2007;**15**(2):577–585.

[13] Fant, G., Liljencrants, J. and Lin, Q.G. A four-parameter model of glottal flow. *Speech Transmission Laboratories QPSR* 1985;**4**:1–13.

[14] Markel, J.D. and Gray, A.H. *Linear Prediction of Speech*. Springer-Verlag, Berlin, 1976.

[15] Rossiter, D.P., Howard, D.M. and Downes, M. A real-time LPC-based vocal tract area display for voice development. *Journal of Voice* 1995;**8**(4):314–319.

[16] Rossiter, D. and Howard, D.M. ALBERT: A real-time visual feedback computer tool for professional vocal development. *Journal of Voice* 1996;**10**(4):321–336.

[17] Howard, D.M., Welch, G.F., Brereton, J., Himonides, E., DeCosta, M., Williams, J. and Howard, A.W. WinSingad: A real-time display for the singing studio. *Logopedics Phoniatrics Vocology* 2004;**29**(3):135–144.

[18] Howard, D.M., Brereton, J., Welch, G.F., Himonides, E., DeCosta, M., William, J. and Howard, A.W. Are real-time displays of benefit in the singing studio? An exploratory study. *Journal of Voice* 2007: **21**(1):20–34.

[19] Schroeter, J. and Sondhi, M.M. Techniques for estimating vocal-tract shapes from the speech signal. *IEEE Transactions on Speech and Audio Processing* 1994;**2**(1)II:133–150.

[20] Abberton, E.R.M., Howard, D.M. and Fourcin, A.J. Laryngographic assessment of normal voice: a tutorial. *Clinical Linguistics and Phonetics* 1989;**3**(3):281–296.

[21] Cooper, C., Howard, D.M., Tyrrell, A.M. and Murphy, D. Singing synthesis with an evolved waveguide mesh model. *IEEE Transactions on Speech and Audio Processing* 2006;**14**(4):1454–1461.

[22] Koza, J. *Genetic Programming: On the Programming of Computers by Means of Natural Selection*. MIT Press, Boston, 1992.

[23] Holland, J.H. *Adaption in Natural and Artificial Systems*. University of Michigan Press, 1975.

[24] Lones, M.A. and Tyrrell, A.M. Modelling biological evolvability: implicit context and variation filtering in enzyme genetic programming. *BioSystems* 2004;**76**(1–3): 229–238.

[25] Cooper, C., Howard, D.M. and Tyrrell, A.M. Using GAs to create a waveguide model of the oral vocal tract. *Lecture Notes in Computer Science* 2004;**3005**:280–288.

[26] Howard, D.M., Tyrrell, A.M., Murphy, D.T., Cooper, C. and Mullen, J. Bio-inspired evolutionary oral tract shape modelling for physical modelling vocal synthesis. *Journal of Voice* 2009;**23**(1):11–20.

[27] Howard, D.M. and Murphy, D.T. *Voice Science Acoustics and Recording*. Plural Press, San Diego, 2008.

6.3

How Genetic Algorithms can Improve Pacemaker Efficiency

Laurent Dumas[1] and Linda El Alaoui[1,2]

[1]*Laboratoire Jacques-Louis Lions, Université Pierre et Marie Curie, Paris, France*

[2]*Département de Mathématiques, Institut Galilée, Villetaneuse, France*

6.3.1 Introduction

Every year throughout the world, millions of people die from cardiovascular diseases and in certain regions they are the main cause of death. Medical research enables a better comprehension of the cardiovascular system and provides new adapted treatments. Nevertheless, this research requires large-scale and time-consuming experiments, involving invasive techniques. Numerical simulations have proven to be a useful tool that can help medical researchers. In this work we are interested in the simulated treatment of some diseases caused by a dysfunction of the heart. The heart is located between the lungs and constitutes four parts, the right and left atria and ventricles. The function of the heart is to pump blood from the lungs to the body, allowing the organs to function. The pumping action is the result

Genetic and Evolutionary Computation: Medical Applications   Edited by Stephen L. Smith and Stefano Cagnoni
© 2011 John Wiley & Sons, Ltd

of a contraction/relaxation process induced by an electrical impulse moving across the heart. The electrical signal is first induced in the sinus node, the natural pacemaker, then propagates through the atria and reaches the ventricles through the atrioventricular node. In the ventricles, the propagation is led by the bundle of His, causing a wavefront which propagates by a cell-to-cell activation.

In each cell, a depolarization phase occurs corresponding to the inflow of sodium ions (causing the electrical activation) followed by a plateau phase, and then by a repolarization phase corresponding to the outflow of potassium ions. The electrical activity is generally modeled by the so-called bidomain equations (Henriquez, 1993), in which the current term due to ionic exchanges can be modeled either by a physiological model or by a phenomenological model. The electrical conduction of the heart may be defective, causing the heart to beat too fast, too slow or in an irregular manner. In some pathologies, for example, sinus node dysfunction or bundle branch block are treated with an artificial pacemaker which is used to help the heart recover a quasi-normal electrical activity. A pacemaker consists of a small battery and electrodes transmitting the electrical impulse.

Though today pacemakers give good results, certain questions still arise. How many electrodes should be set? Where should the electrodes be placed? When should the electrodes act? Many studies set out to answer these questions (see Penicka *et al.*, 2004 and references cited therein).

Our aim in this work is to determine the optimal position of the electrodes of a pacemaker on a diseased heart. This can be interpreted as an inverse-type optimization problem which can be solved with optimization tools such as genetic algorithms (GAs). GAs are already used in many other medical applications, for instance in cardiology, the classification of ischaemic beats (Goletsis *et al.*, 2004). GAs are well adapted when the cost function is not smooth or needs a complex simulation. It is well known that the drawback of GAs is the computational cost. In our work, we circumvent this high cost by introducing surrogate models, allowing us to construct an approximation of the cost function. We validate our results by deriving numerical electrocardiograms (ECGs) that are the interpretation of the electrical activity of the heart over time. The ECG is commonly used to diagnose possible heart disease. A numerical study of ECGs has been done (Boulakia *et al.*, 2007).

This chapter is organized as follows. In Section 6.3.2 we present the bidomain/Mitchell-Schaeffer model used to perform the numerical simulation of the cardiac electrical activity. Section 6.3.3 is devoted to the optimization description, and in Section 6.3.4 we present and discuss some numerical results on a simplified test case representative of a left bundle branch block in a modeled heart.

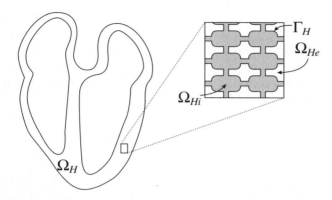

Figure 6.3.1 Simplified view of the heart at macro/microscopic level

6.3.2 Modeling of the Electrical Activity of the Heart

At the microscopic level, the cardiac muscle, denoted by Ω_H, is made up of two distinct and intricate media: the intra- and extracellular media, respectively called Ω_{Hi} and Ω_{He}, which are separated by a surface membrane Γ (see Figure 6.3.1). We denote by ϕ_i and ϕ_e the intra- and extra-cellular potential, then we define the transmembrane potential V_m by

$$V_m(t, x) = \phi_i(t, x) - \phi_e(t, x), \quad \forall(t, x) \in (0, T] \times \Omega_H, \qquad (6.3.1)$$

with $T > 0$. We also introduce the σ_i and σ_e intra- and extracellular conductivity tensors. Thus, the so-called bidomain model consists in the following two degenerate parabolic reaction–diffusion partial derivative equations (PDEs) coupled to an ordinary differential equation (ODE) (Henriquez, 1993):

$$\begin{cases} C_m \partial_t V_m + I_{ion}(V_m, w) - \mathrm{div}(\sigma_i \nabla V_m) = I_{app}, & \text{in } \Omega_H \times (0, T), \\ C_m \partial_t V_m + I_{ion}(V_m, w) + \mathrm{div}(\sigma_e \nabla V_m) = I_{app}, & \text{in } \Omega_H \times (0, T), \\ \partial_t w + g(V_m, w) = 0, & \text{in } \Omega_H \times (0, T). \end{cases} \qquad (6.3.2)$$

Here, C_m is the membrane capacitance and I_{app} an external applied volume current. The above PDEs describe the evolution of the averaged potential ϕ_i and ϕ_e, whereas the last equation in (6.3.2) describes the electrical behavior of the myocardium cell membranes. This ODE is known as the ionic model. The choice of the ionic model defines the nonlinear term $I_{ion}(V_m, w)$ and the function $g(V_m, w)$. On assuming that the heart is electrically isolated, the following boundary condition on the heart boundary $\partial\Omega_H$ holds:

$$\sigma_i \nabla\phi_i \cdot n = \sigma_e \nabla\phi_e \cdot n = 0, \tag{6.3.3}$$

where n denotes the outward unit normal to $\partial\Omega_H$. We complete the system (6.3.2)–(6.3.3) by the initial condition

$$V_m(0, x) = V_m^0(x) \quad \text{in } \Omega_H. \tag{6.3.4}$$

The current term due to ionic exchanges, I_{ion}, is calculated here by solving the physiological model of Mitchell and Schaeffer (2003). This model has been chosen as it allows numerically realistic ECGs to be obtained, see Boulakia *et al.* (2007) for more details:

$$I_{ion} = -\frac{w}{\tau_1} V_m^2(1 - V_m) + \frac{V_m}{\tau_2}, \tag{6.3.5}$$

where the auxiliary variable w satisfies the following ODE:

$$\frac{dw}{dt} = g(V_m, w), \quad \text{with} \quad g(V_m, w) = \begin{cases} \dfrac{w - 1}{\tau_3} & \text{if } V_m < V_g, \\[2mm] \dfrac{w}{\tau_4} & \text{if } V_m > V_g, \end{cases} \tag{6.3.6}$$

and τ_1, τ_2, τ_2, τ_3, τ_4 and $V_g < 1$ are given positive constants.

Note that in this model few parameters have to be fitted, which is an advantage from a numerical point of view. We refer to Boulakia *et al.* (2008) for a mathematical analysis of the bidomain equations coupled with the Mitchell and Schaeffer model.

Remarks:

1. In this work we shall validate our numerical optimization strategy by deriving ECGs. Usually, for deriving ECGs from the bidomain model only the intracellular domain is assumed to be electrically isolated and the boundary condition on the extracellular potential u_e involves the surrounding tissue. We refer to Boulakia *et al.* (2007) for the description of the system coupling the heart and the torso.
2. For modeling the electrode of a pacemaker in the bidomain model we add a local volumic source term in the PDEs of system (6.3.2).

6.3.3 The Optimization Principles

6.3.3.1 The Cost Function

Here, we investigated various cost functions to optimize the positioning of the electrodes of a pacemaker on a diseased heart.

The first cost function that has been tested is the quadratic norm in space and time of the difference between the transmembrane potential V_m of a diseased heart with a given position of electrodes and its target value $V_{m,target}$ computed for the healthy case:

$$J_1 = \int_0^T \int_{\Omega_H} |V_m - V_{m,target}|^2 dx dt. \tag{6.3.7}$$

Actually, this first and natural cost function does not give satisfactory results for two reasons. First, it is due to the fact that the electrical activity of electrodes will represent a major obstacle to make V_m converge to $V_{m,target}$ on the whole domain in space and time. A second reason is that the correct criteria to recover a normal electrical activity is rather to reduce the delay of a characteristic depolarization time.

A new and better cost function is thus introduced and is expressed as

$$J_2 = t_d - t_{d,target}, \tag{6.3.8}$$

where t_d represents the first time for which 95% of the whole heart is depolarized:

$$t_d = \inf\{t \geq 0, \quad \text{Volume}(\Omega_{Ht}) \geq 0.95\,\text{Volume}(\Omega_H)\},$$

with

$$\Omega_{Ht} = \{x \in \Omega_H, \quad V_m(t, x) > V_s\}.$$

As previously, $t_{d,target}$ denotes the same value for the corresponding healthy heart.

6.3.3.2 The Optimization Algorithm

All the cost functions previously described are computed after solving a complex set of coupled PDEs and ODE with strong three-dimensional effects. Moreover, due to the complexity of the heart geometry, they display a non-smooth behavior with many local minima. For all these reasons, the minimization of these cost functions is achieved by using evolutionary algorithms and more precisely genetic algorithms.

At the first stage of this work (Dumas and El Alaoui, 2007), a classical real-coded genetic algorithm has been used to optimize the positioning of one or two electrodes of a pacemaker on the internal boundary surface of the heart. The selection process used in the genetic algorithm is done with a proportionate roulette wheel with respective parts based on the rank of each element in the population. The crossover of two elements is obtained by a barycentric combination with random and independent coefficients in each coordinate, whereas the mutation of one element is of non-uniform type. Finally, a one-elitism principle is added to ensure the best element of the previous generation is preserved.

More recently, in order to improve the convergence speed of this algorithm, the use of a genetic algorithm with a surrogate model described in the next paragraph has been investigated.

6.3.3.3 A New Genetic Algorithm with a Surrogate Model

A classical speed up of the convergence of a genetic algorithm when the computational time of the cost function $x \mapsto J(x)$ is high, takes advantage of the large and growing data base of exact evaluations by evaluating an approximation $x \mapsto \tilde{J}(x)$ of the exact function at a lower cost. In the literature this technique is usually called surrogate or meta-model: Giannakoglou (2000), Ong *et al.* (2003), Jin (2005), Abou *et al.* (2008). In the present work, the chosen strategy consists of performing exact evaluations only for all the best fitted elements of the population. The new algorithm, called an Approximate Genetic Algorithm (AGA), is thus deduced from a classical genetic algorithm by changing the evaluation phase for a given generation n_g into the following:

- *Compute the approximated cost function $x \mapsto \tilde{J}(x)$ for all elements x of the population.*
- *Evaluate exactly the best $N_b(n_g)$ elements in respect of $\tilde{J}$. In other words, J is computed for those $N_b(n_g)$ elements.*

The shape of the function $n_g \to \frac{N_b(n_g)}{N_{pop}}$ representing the rate of exact evaluations at generation n_g is chosen first. In any case, it has to be equal to 1 for the first $N_{geninit}$ generations. Then, this ratio is linearly decreasing from a value $r \in (0, 1)$ to 0.

The interpolation method chosen here comes from the field of neural networks and is called RBF (Radial Basis Function) interpolation (Giannakoglou, 2000). Suppose that the function J is known on N points $\{T_i, 1 \le i \le N\}$.

The idea is to approximate J at a new point x by a linear combination of radial functions:

$$\tilde{f}(x) = \sum_{i=1}^{n_c} \psi_i \Phi(||x - \hat{T}_i||), \qquad (6.3.9)$$

where

- $\{\hat{T}_i, 1 \leq i \leq n_c\} \subset \{T_i, 1 \leq i \leq N\}$ is the set of $n_c \leq N$ nearest points to x for the Euclidian norm $||.||$, on which an exact evaluation of J is known.
- Φ is a radial basis function chosen in the following set:

$$\Phi_1(u) = \exp\left(-\frac{u^2}{r^2}\right),$$

$$\Phi_2(u) = \sqrt{u^2 + r^2},$$

$$\Phi_3(u) = \frac{1}{\sqrt{u^2 + r^2}},$$

$$\Phi_4(u) = \exp\left(-\frac{u}{r}\right),$$

for which the parameter $r > 0$ is the attenuation parameter.

The scalar coefficients $(\psi_i)_{1 \leq i \leq n_c}$ are obtained by solving the least-square problem of size $N \times n_c$

$$\text{minimize} \quad err(x) = \sum_{i=1}^{N} (J(T_i) - \tilde{f}(T_i))^2.$$

6.3.3.4 Results of AGA on Test Functions

The AGA method has been tested on various analytical test functions, and among them the well-known Rastrigin function, with n variables and of parameter a:

$$Rast_a(x_1, \ldots, x_n) = \sum_{i=1}^{n} (x_i^2 - a \cos(2\pi x_i)) + an,$$

defined on $\mathcal{O} = [-5, 5]^n$ for which there exist many local minima and only a global minimum located at $x^* = (0, \ldots, 0)$. Note that the number of local minima in $\mathcal{O}$ is always equal to 11^n but the attraction basins become steeper when $a > 0$ takes a higher value.

A statistical comparison, based on 40 independent runs for each method, has been done between a classical genetic algorithm and its associated AGA

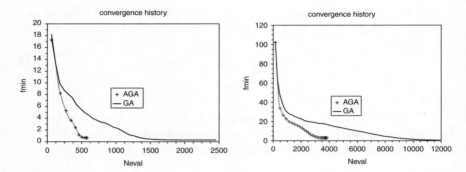

Figure 6.3.2 Comparison of the convergence speed of GA and AGA on $Rast_1$, n = 6 (left) and n = 20 (right)

method. For each method, the average best obtained value is plotted with respect to the number of exact function evaluations needed to achieve it.

The results obtained by the AGA method and the associated GA are compared in the case of the $Rast_1$ function, with 6 and 20 parameters respectively (see Figure 6.3.2). The population size for the GA is either equal to $N_p = 40$ if $n = 6$ or $N_p = 160$ if $n = 20$. A maximal number of function evaluations is allowed, either equal to 2500 when $n = 6$ or 12,000 when $n = 20$.

In both simulations, the specific parameters associated with the AGA method are respectively equal to $N_{geninit} = 2$ (number of initial generations with only exact evaluations) and $n_c = 16$ (size of the sampling to compute the approximated function).

The AGA overperforms the GA method in terms of exact evaluations of the cost function to achieve a given level of convergence. A reduction rate approximately equal to 3 is observed here in the case of the Rastrigin function.

Compared with other results found in the literature for the Rastrigin function, the results obtained with the AGA method are very competitive. Moreover, it is worth noticing that the same general conclusions have been observed for other test functions similar to the Rastrigin function, that is with many local minima, namely the Ackley and the Griewank functions.

6.3.4 A Simplified Test Case for a Pacemaker Optimization

6.3.4.1 Description of the Test Case

The simulations are performed on a simplified geometry which contains ventricles only, see Figure 6.3.3.

Figure 6.3.3 A simplified heart geometry Ω_H

The domain, closed to a human heart, is analytically defined as the union of four truncated ellipsoids:

$$\left(\frac{x}{a_{iL}}\right)^2 + \left(\frac{y}{b_{iL}}\right)^2 + \left(\frac{z}{c_{iL}}\right)^2 = 1, \quad \left(\frac{x}{a_L}\right)^2 + \left(\frac{y}{b_L}\right)^2 + \left(\frac{z}{c_L}\right)^2 = 1,$$

with $\{a_{iL}, b_{iL}, c_{iL}, a_L, b_L, c_L\} = \{2.72, 2.72, 5.92, 4, 4, 7.2\}$ cm for the left ventricle internal and external boundary respectively, and

$$\left(\frac{x}{a_{iR}}\right)^2 + \left(\frac{y}{b_{iR}}\right)^2 + \left(\frac{z}{c_{iR}}\right)^2 = 1, \quad \left(\frac{x}{a_R}\right)^2 + \left(\frac{y}{b_R}\right)^2 + \left(\frac{z}{c_R}\right)^2 = 1,$$

with $\{a_{iR}, b_{iR}, c_{iR}, a_R, b_R, c_R\} = \{7.36, 3.36, 6.2, 8, 4, 6.84\}$ cm for the right ventricle. All these ellipsoids are restricted to the half space $z \leq 2.75$.

Note that the pathology considered here is a left bundle branch block, for which only the right ventricle is initially stimulated. In this case, the electrodes can be placed in the atria and/or in the ventricles. As we only simulate here the ventricles, we seek for the optimal position of the electrodes in the internal surface of the left ventricle, the endocardium. A mapping from the endocardium or a part of it to a simple plane domain, for instance a rectangular domain of $\mathbb{R}^2$, has first been defined in order to simplify the parametric search space. Note that other pathologies, like a right bundle branch block, have also been investigated with the same method but are not presented here as the conclusions are very similar. In particular, the same cost function J_2 defined in (6.3.8) has been used for the study of all these pathologies.

Figure 6.3.4 Transmembrane potential in the ventricles at $t = 28.5$ ms in the healthy, pathologic and pathologic case treated with one electrode (from left to right)

6.3.4.2 Numerical Results

We choose the conductivities in (6.3.2) such that the anisotropy of the fibers in the myocardium are taken into account, namely $\sigma_i = \alpha_i^t(I - d_f \otimes d_f) + \alpha_i^l(I - d_f \otimes d_f)$ and $\sigma_e = \alpha_e^t(I - d_f \otimes d_f) + \alpha_e^l(I - d_f \otimes d_f)$, where d_f is the direction of the fibers, I the identity matrix in $\mathbb{R}^3$, and $\alpha_i^t = 5 \ 10^{-3}$, $\alpha_i^l = 1.5 \ 10^{-1}$, $\alpha_e^l = 1. \ 10^{-1}$ and $\alpha_e^t = 7.5 \ 10^{-3}$. The parameters in (6.3.2)–(6.3.6) are chosen as follows: $C_m = 1$, $\tau_1 = 0.8$, $\tau_2 = 18$, $\tau_3 = 300$, $\tau_4 = 100$, $V_g = -67$. The domain Ω_H is discretized with tetrahedra for a total number of nodes approximately equal to 80,000. The bidomain problem (6.3.2)–(6.3.6) is approximated by a piecewise finite elements scheme in space and by a second-order backward differences scheme in time. The simulations are done with the C++ library *LifeV*[1].

The intensity of the initial stimulation equals 0.5 mV during 10 ms. The artificial stimulations have the same intensity as the initial stimulation and hold during 10 ms. As we are interested in the depolarization phase only, the final time of computations is only equal to 100 ms, whereas the total duration of the depolarization–repolarization process is approximately 600 ms. An example of the propagation of the depolarization front is given in Figure 6.3.4 at time $t = 28.5$ ms for the three following cases: a healthy heart, a pathologic heart and a pathologic heart treated with one electrode.

The GA population consists of 30 individuals for optimization of a single electrode position and 60 individuals for two electrodes. A near optimal solution is achieved after 10 generations. However, it should be noted that the maximum number of evaluations never exceeds 100 because of the use of the AGA algorithm.

---

[1] http://www.lifev.org/

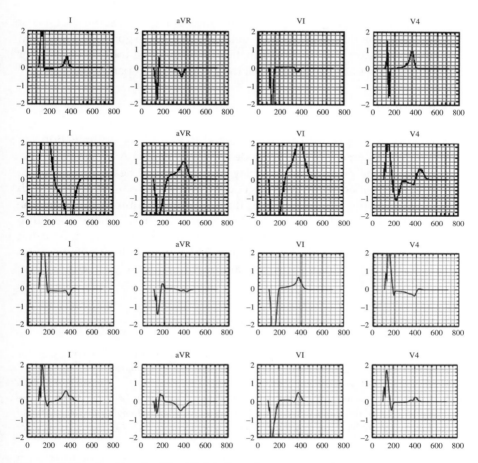

Figure 6.3.5 Four leads of the ECG in the reference case, the pathologic case, the pathologic case with one or two electrodes (from top to bottom)

We present the results obtained by using the cost function defined in (6.3.8). In this case, the isolated model (6.3.2)–(6.3.6) is used to simulate the electrical activity of the heart. In the pathological case, $J_2 = 73$ ms whereas this delay is reduced to $J_2 = 34.75$ ms and $J_2 = 20.75$ ms when one, and respectively two, electrodes were placed at the obtained optimal positions. Note also that other positions of one electrode can lead to a delay of up to 65 ms. Such a high value shows the interest of an optimization process.

Figure 6.3.5 shows four leads of the ECG used in the reference case, the pathologic case and the pathologic case with one and two, electrodes after an optimization with J_2. It is particularly interesting to observe that a cost function based on the depolarization delay also leads to a good ECG recovery. It seems to indicate that this cost function is definitely a robust criterion for the optimization of the positioning of the electrodes of a pacemaker.

6.3.5 Conclusion

In this work, a simple test case is presented to show how optimization can improve the placement of the electrodes of a pacemaker. A robust cost function based on the depolarization delay of the diseased heart is introduced. The results obtained show how a good ECG can be recovered, even with only one electrode, after optimizing its positioning. In order to reduce the computational time of this optimization process, a new genetic algorithm coupled with a surrogate model has also been introduced. Indeed, by using a more realistic model of the heart geometry of a given patient, the medium-term objective is to help practitioners choose the best strategy to cure a given heart pathology.

References

Abou El Majd, B., Desideri, J.-A. and Duvigneau, R. 2008 Multilevel strategies for parametric shape optimization in aerodynamics. *European Journal of Computational Mechanics* **17**, 149–168.

Boulakia, M., Fernández, M.A., Gerbeau, J.-F. and Zemzemi, N. 2007 Towards the numerical simulation of electrocardiograms. In *Functional Imaging and Modeling of the Heart. Lecture Notes in Computer Science*, Vol. 466, pp. 240–249. Springer-Verlag, Berlin.

Boulakia, M., Fernández, M.A., Gerbeau, J.-F. and Zemzemi, N. 2008 A coupled system of PDEs and ODEs arising in electrocardiograms modeling. *Applied Mathematics Research Express* 2008: abn002-28.

Dumas, L. and El Alaoui, L. 2007 How genetic algorithms can improve a pacemaker efficiency. Proceedings of GECCO 2007, pp. 2681–2686.

Giannakoglou, K.C. 2000 Acceleration of GA using neural networks, theoretical background, GA for optimization in aeronautics and turbomachinery. *VKI Lecture Series*.

Goletsis, Y., Papaloukas, C., Fotiadis, D.I., Likas, A. and Michalis, L.K. 2004 Automated ischemic beat classification using genetic algorithms and multicriteria decision analysis. *IEEE Transactions on Biomedical Engineering* **51** (10), 1717–1725.

Henriquez, C.S. 1993 Simulating the electrical behavior of cardiac tissue using the bidomain model. *Critical Reviews in Biomedical Engineering* **21** (1), 1–77.

Jin, Y. 2005 A survey on fitness approximation in evolutionary computation. *Journal of Soft Computing* **9**, 3–12.

Mitchell, C.C. and Schaeffer, D.G. 2003 A two-current model for the dynamics of cardiac membrane. *Bulletin of Mathematical Biology* **65**, 767–793.

Ong, Y.S., Nair, P.B. and Keane, A.J. 2003 Evolutionary optimization of computationally expensive problems via surrogate modeling. *AIAA Journal* **41**, 687–696.

Penicka, M., Bartuneck, J., De Bruyne, B., Vanderheyden, M., Goethals, M., De Zutter, M., Brugada, P. and Geelen, P. 2004 Improvement of left ventricular function after cardiac resynchronisation therapy is predicted by tissue Doppler imaging echocardiography. *Circulation* **109** (8), 978–983.

7

The Future for Genetic and Evolutionary Computation in Medicine: Opportunities, Challenges and Rewards

It is apparent from the number of papers published in journals and presented at international conferences that there is an increasing interest in the use of GEC in medical applications. One of the great attractions of GEC is the flexible way in which it can be implemented, either in conjunction with other methods such as fuzzy systems and neural networks, or standalone. Many variants and combinations of GEC methods are also continually being developed, each with its own advantages and disadvantages, depending on the nature of the application and the data to be processed.

One fundamental difference in its application has been the specific role that GEC plays, for example, as an optimiser of other computational intelligence techniques or in the direct processing of the data. There is also scope in many applications for GEC to be applied to raw data, with little or no a priori knowledge of the nature of this data. This offers great opportunities but also some challenges, as outlined below.

Genetic and Evolutionary Computation: Medical Applications   Edited by Stephen L. Smith and Stefano Cagnoni
© 2011 John Wiley & Sons, Ltd

7.1 Opportunities

One of the great opportunities of applying GEC to the medical and health-related sector is the huge range of applications readily available to benefit from the technology. The medical community has been slow to encompass technology compared with other sectors, for a number of reasons. Firstly, due to the nature of the sector, which has responsibility for patients' treatment and well-being, there has always been a reluctance to adopt new techniques when training of health professionals is strongly grounded in following previous experience and procedures. Secondly, there has been, until relatively recently, limited funding to explore new opportunities from health. Finally, obtaining ethical approval and other statutory regulation can be a daunting and time-consuming task.

Computer literacy of health professionals (and the population in general) is increasing as greater use of computing systems is required in everyday life. This has overcome some of the natural reluctance of health professionals to allow the computer into the clinic. Similarly, patients of all ages can see the potential benefit of introducing technology into the monitoring, diagnosis and treatment activities and probably wonder why so much of medical practice is still reliant on highly subjective practices.

Another motivation for the introduction of technology is cost driven. People are living longer and the population is getting steadily greater, particularly the elderly who will require more healthcare. All health services are looking for ways to be more efficient in delivering healthcare and hence reduce costs.

7.2 Challenges

Although there may be many opportunities in applying GEC to medicine, there are also challenges to overcome.

The first hurdle is to establish the clinical need.

How can the technology provide benefits to the patient and the health service?

If this question cannot be satisfactorily addressed in terms such as increased quality of life or life expectancy and overall reduced costs, then it is unlikely to succeed beyond a proof-of-concept trial.

Will the resulting solution meet the expectations and requirements of the health professionals?

One of the greatest challenges is ensuring that the correct problem is being addressed. As with all technology-based projects there is often a difference in understanding between the client and the technology provider. This can be particularly true with health professionals who are unlikely to have a strong grasp of what computational intelligence can be expected to deliver. It is therefore essential that time is taken to explore the short, medium and long-term expectations of the technology and for both parties to commit to a clearly written specification.

Is there a need to support existing diagnostic practice rather than replace it?

The intended role of the GEC technology in diagnosis is critical. Fully automated diagnosis is both ambitious and costly to implement, requiring extensive evaluation to ensure it is both accurate and safe in all foreseeable situations. The use of such technology to support a health professional's diagnosis encompassing other factors is a more acceptable proposition both to the health profession as well as its patients.

How does the technology fit into current clinical practice?

An issue often overlooked until it's too late, is how easily the proposed solution will fit into the current clinical Doctors' surgeries, hospitals and other healthcare environments are busy places, the organisation of which has been established over many years. Introducing even what may be considered by the investigator to be a modest change to the clinical practice can have considerable impact on costs, time, workloads and staff morale. Have the costs of the new technology been fully evaluated in terms of procurement, consumables, maintenance and decommissioning? Are there impacts on the roles and workloads of staff, for example a nurse testing patients and consultants evaluating results?

Your algorithm is only as good as the data it is trained on!

Obtaining good data with which to train and test your algorithms is one of the greatest challenges in applying GEC to medical applications. For the system to be successful the data has to be both correct and representative of that to be encountered in deployment. For many areas of medical application considered in this book, there is rarely a way of obtaining a 'gold standard' or 'ground truth' that is 100% accurate. The paradox is that if such data was easily obtainable then the technology would be unlikely to be of any greater benefit to the medical community. Many, if not the majority of clinical diagnoses are at some point dependent upon subjective assessment, which is difficult to undertake and, thus, can be unreliable. This also has the consequence that a substantial number of data examples are required for training and test sets. In the case of patient studies, this has added time and cost implications.

Applications for ethical approval will often require statistical justification for the patient sample size, to avoid situations where either the number of patients measured is too small to give meaningful results or larger than is necessary, with the effect of causing pain, distress or inconvenience where it isn't warranted.

A second major consideration is that evolutionary algorithms can be trained to discriminate between classes in a training set, but how sure can you be that it is the medical condition of interest which is the discriminating factor? If all the patients with the condition of interest are measured in one context, such as a specific location, time of day or environment, and the controls in another, it may be the context that has led to the classification, not the medical condition of interest. For example, if all breast cancer patients' mammograms were scanned with one manufacturer's equipment and all controls' mammograms scanned with another, the algorithm may be discriminating on some artefact of the digitising process, rather than the presence or absence of cancer.

Is GEC the most appropriate technology?

Finally, it is important to recognise when GEC is likely to be an appropriate technology to apply rather than more traditional technologies. A good reason for applying GEC may be that the traditional signal-processing techniques have often been found to be unsuccessful or unreliable, maybe because the nature of the problem is highly non-linear. However, traditional techniques, if appropriate, are likely to be simpler, easier to understand and more efficient to apply, where implementation or statutory regulation is concerned.

7.3 Rewards

Having considered the opportunities and challenges of using GEC for medical applications, what long-term benefits can be expected? The combination of low take-up of computational intelligence in medical practice and the unique benefits offered by GEC points to the following rewards:

- improved diagnosis, treatment and therapy for patients;
- cost savings for healthcare providers, patients and carers;
- commercial opportunities.

The primary beneficiary must be the patient or, in the case of preventative medicine, the general public. However, in order to deliver this technology there must also be overall cost savings for the healthcare provider to motivate and commit to adoption. Knowledge transfer and, specifically, commercialisation is the key to achieving this adoption as it is only the promise of profit and return that can generate the investment required to see any new technology introduced to routine healthcare.

7.4 The Future for Genetic and Evolutionary Computation in Medicine

The conclusion we are left with at the end of this book is the exciting realisation that medical applications of GEC will become more prevalent and successful in delivering better healthcare to patients, better support for health professionals and a valuable technology for the healthcare industries. For this development to be effective and sustainable, it is essential that appropriate GEC methodologies are adopted for the right applications. This isn't always easy, especially for those new to GEC, but there is a strong GEC community which is both supportive and knowledgeable. This community is easily engaged with, through online newsgroups, tutorials, workshops, paper presentations and networking opportunities at conferences and by reading the growing literature. (A list of some of the more prominent sources is presented in the appendix.) Undoubtedly, there has never been a better time to exploit the rich and varied source of medical applications which will confirm genetic and evolutionary computation as a central technology in medical engineering.

Appendix

Introductory Books and Useful Links

Introductory Books

- A.E. Eiben, J.E. Smith, *Introduction to Evolutionary Computing*, Springer, 2003.
- W. Banzhaf, P. Nordin, R.E. Keller, F.D. Francone, *Genetic programming, An Introduction*, Morgan Kaufmann, 1998.
- C. Blum, D. Merkle (eds) *Swarm Intelligence: Introduction and Applications*, Springer, 2008.

Conferences

- **http://www.sigevo.org/**
  GECCO: Genetic and Evolutionary Computation Conference.
- **http://ieee-cis.org/**
  IEEE WCCI: World Conference on Computational Intelligence. IEEE CEC: Conference on Evolutionary Computation, is incorporated in this joint event.
- **http://www.evostar.org**
  Evostar: EuroGP/EvoCOP/EvoBio/EvoApplications; the major European set of joint events on EC.

Genetic and Evolutionary Computation: Medical Applications   Edited by Stephen L. Smith and Stefano Cagnoni
© 2011 John Wiley & Sons, Ltd

- **http://ls11-www.cs.uni-dortmund.de/ppsn/**
  PPSN (Parallel Problem Solving from Nature): a peculiar and scientifically
  very relevant all-poster conference, dedicated to EC and other computing
  paradigms gleaned from natural processes.

Journals

- **http://www.mitpressjournals.org/loi/evco/**
  Evolutionary Computation.
- **http://ieee-cis.org/pubs/tec/**
  IEEE Transactions on Evolutionary Computation.
- **http://www.springer.com/computer/artificial/journal/10710**
  Genetic Programming and Evolvable Machines.

Projects, SIGs, etc.

- **http://www.sigevo.org**
  The ACM Special Interest Group on Evolutionary Computation.
- **http://www.cs.bham.ac.uk/~wbl/biblio**
  The Genetic Programming Bibliography. A very extensive searchable collection of references on Genetic Programming.

Index

Genetic and Evolutionary Computation: Medical Applications   Edited by Stephen L. Smith and Stefano Cagnoni
© 2011 John Wiley & Sons, Ltd